POSTHARVEST TECHNOLOGY OF KINNOW

Dr. A. D. Sharma
Assistant Professor
Guru Angand Dev Veterinary and
Animal Sciences University,
Ludhiana (Punjab)

SCIENTIFIC PUBLISHERS (INDIA)
P.O. BOX 91 JODHPUR

Published by:
Tanay Sharma
Scientific Publishers (India)
5-A, New Pali Road, P.O. Box 91
Jodhpur - 342 001

E-mail: info@scientificpub.com
www.scientificpub.com

ISBN: 978-81-7233-586-1

Lasertype set: Rajesh Ojha
Printed in India

"To my beloved wife and son"

FOREWORD

In India, substantial postharvest losses occur in fruits and vegetables. A reduction in these losses would lead to increased revenue and better nutrition. This publication provides some useful insights into postharvest practices used by Punjab growers and into growers' awareness about available scientific methods of postharvest handling of fruits and vegetables. The author has divided the subject matter into five chapters with appropriate headings. The introduction covers the status, and the rationale for doctoral research conducted by the author. The next chapter covers various techniques of enhancing shelf life of fruits and vegetables. The third chapter deals with mechanical handling of fruits and vegetables. The development and performance evaluation of developed prototype for cleaning, waxing, and grading of Kinnow is covered in the fourth chapter. The knowledge regarding the low-cost prototype for cleaning, waxing, and grading of Kinnow should prove beneficial for growers and entrepreneurs involved in postharvest handling of fruits. Another important aspect, i.e., quality characteristics of fruits, is covered in the last chapter that includes recent developments in the field of non-destructive methods of quality determination of fruits and vegetables.

This review publication should help in the dissemination of technology and knowledge to those who do not have access to relevant research journals. This book will not only be useful for researchers, but also for students and teachers in the field of postharvest handling of fruits and vegetables. The book covers almost all of the essential aspects of postharvest handling of fruits and vegetables.

I trust that this publication would serve as a source of information, for the intended audience, on postharvest handling of fruits and vegetables in Punjab as well as in rest of India.

Dr. Manjit S. Kang
Vice Chancellor
Punjab Agricultural University
Ludhiana 141 004

ACKNOWLEDGEMENT

The author wishes to express his sincere gratitude to Dr. B.S. Ghuman, Senior Research Engineer and former Head, Department of Processing and Food Engineering, Punjab Agricultural University, Ludhiana for his scientific temperament, disciplined way of working, valuable guidance, continued encouragement, constructive suggestions and providing the necessary facilities which led to the successful completion of authors Ph.D. dissertation.

Author wishes to thanks Er. B.S. Sidhu, Director, Department of Agriculture, Punjab, Er. Aishvarya Sharma, Er. Jagdish Singh, and Er. Harcharan Singh, all Assistant Agriculture Engineer, Department of Agriculture, Punjab, for their valuable help throughout the study.

The author wishes to acknowledge the financial help given by the University Grants Commission in the form of Junior Research Fellowship, without which it would be difficult for the author to complete his doctoral degree.

Sincere and special thanks are due towards Mr. K.N. Narula, Assistant Librarian, M.S. Randhawa Library, Punjab Agricultural University Ludhiana who encouraged me and give his valuable time and suggestions to bring out my research work in the form of this publication.

Words fail the author to express his deep sense of gratitude and appreciation towards his family members for their inspiration, moral and financial support and encouragement without which the author would not have been able to bring out this publication.

Dr. Amandeep Sharma

PREFACE

Fruits are good source of nutrition as they provide vitamins and minerals for proper growth of human beings. In our country, the postharvest losses in fruits are about 30 per cent. Such a huge losses are not only leading to less availability of nutritious food but also cause economic loss to the country and growers. Kinnow is an important fruit of Punjab state. Its proper handling in scientific manner is the need of the hour. Owing to its good taste juice content, this fruit has export potential.

The work on development of prototype for on-farm pre processing of Kinnow was taken to help the growers to minimise the postharvest losses and to add value at farm itself. This will help in reducing the number of handlings the fruit has to undergo prior to processing at the processing centres.

I took this opportunity to publish my Ph.D. dissertation work for the benefit of students, researchers and all those who are concerned with reducing the losses of fruits and vegetables.

This book introduces about Kinnow fruit and its present status of postharvest status in Punjab state and about the awareness of farmers regarding the post-harvest operations of Kinnow fruits. The second chapter contains mainly the review of different techniques for enhancement of she life of fruits and vegetables. The techniques such as controlled atmosphere, individual shrink wrapping and application of film coatings has been stressed upon. The third chapter is about the mechanical handling of fruits and vegetables. It includes the various operation which are carried out in a pack house, such as, dumping, cleaning, degreening, waxing, surface drying and size grading. The fourth chapter deals with the develop-ment and evaluation of prototype. This chapter contains the information regarding the concept building and selection of the objectives of the dissertation work taken. This chapter contains the design and development, material estimation and fabrication methodology and also contain the performance evaluation of various parts of the prototype such as cleaning, waxing, surface drying and grading units. The most important part of quality characteristics can not be over looked in any study, so, the chapter five discusses the various quality parameters such and postharvest diseases, physiological disorders, post harvest insect problems and treatments, physical and chemical quality characterists of Kinnow and the recent developments of non-destructives techniques of quality determination in fruits and vegetables. The bibliography contains all the sources, direct or indirect to acknowledge the help taken from various sources of information. If there is any portion that has been left without reference, this may be taken as if done just our of ignorance.

The author wants to admit that some part of the book may appear repetitive in nature but it was after considering all the technical aspects that the various contents has been placed in a logical sequence.

Over and above, feedback is solicited.

Dr. Amandeep Sharma
e-mail:drsharma.aman@gmail.com

CONTENTS

Chapter 1

INTRODUCTION

The scientific methods of post-harvest handling of fruits and vegetables not only facilitate the proper care of agricultural produce but also help in enhancing the shelf-life of the produce. Post-harvest technologists throughout the world constantly work on the principles that can be used to enhance the shelf-life of agricultural produce. These methods may help to reduce the instances of malnutrition. The reduction of post-harvest losses of agricultural produce is also important to the fact that it costs dear to produce something than to save it. In India, the number of fruit diets is much less than those in the developed countries. Therefore, it requires steps towards the proper handling of fruits and vegetables so that the quantity of produce that is lost during the post-harvest period may have opportunity to reach the common man.

Immediately, after harvesting the agricultural produce encounters at least one mass transfer problem, whether it is moisture loss, oxygen intrusion, aroma loss or gain, or oil migration. Given the range of barrier attributes that coatings can provide, these produce can benefit with improved stability, texture, taste, and aroma. Incorporation of antioxidants and/or antimicrobials that are control released can further enhance produce quality. Many commodities also suffer from breakage and disintegration in manufacture packaging, transportation and use. The higher yield easier handling and improved appearance from integrity-enhancing coatings are additional potential advantages. Finally, consumers often associate product quality with appropriate gloss color and tactic feel properties that coatings can affect. All of these features are of even greater importance because of increased consumer interest in quality, variety, and convenience. At the same time, food producers as well as processors are interested in worldwide markets that demand longer shelf life and in providing products with a minimum of packaging which is ultimately recyclable. Thus, these factors have combined to spark interest in exploring the firm-formation, barrier, mechanical, and sensory properties of proteins and other materials (Krochta).

Kinnow is one of the major fruits grown in Punjab state and is relished by the Indian community. Besides Punjab this citrus fruit is also grown in other parts of India. It is mostly grown in tropical and sub-tropical regions of the world. During the past decade the popularity of Kinnow amongst the Indian community has exceeded that of Nagpur orange. The fruit has good demand within the country as well as in the foreign markets.

Kinnow mandarin is a high quality variety which resulted from a cross between 'King' and 'Willow leaf' (Mediterranean) mandarin done by H.B. Frost in 1915. Plants are medium to large, erect, and symmetrical; dense foliage with a few scattered spines, leaves are broadly lanceolate. The Kinnow fruit is medium oblate, and attains deep orange yellow colour on ripening; rind thin, adherent to carpel's (9 to 10 in number), very juicy and having 12 to 24 seeds. It is considered to be originated in China. In India it was introduced from USA and has been found to perform better under Punjab, Himachal Pradesh, Uttar Pradesh and Maharashtra states. Kinnow was introduced in Punjab by Punjab Agricultural University in 1959 at its Regional Fruit Research Station, Abohar. This mandarin has proved to be high yielding and comparatively disease tolerant as compared to other citrus fruits grown in India.

The citrus fruits are non-climacteric and hence eating quality of fruits does not improve after harvest (Singh *et al.*, 2002). Therefore, these fruits are harvested when their internal quality is at their best. Fruit colour can not be taken into consideration while harvesting, since colour development in citrus fruits depends upon cold nights followed by warm days during maturity. Under semi-arid irrigated conditions of Abohar, the optimum harvesting time is considered between January–February, and for sub-mountainous region of Hoshiarpur the harvesting time is first week of February. During this period the total soluble solids (TSS) content, acidity of the fruits, and the ratio of TSS to acidity are 11°Brix, 0.8%, and 13.70, respectively and it has maximum juice content and palatability. The delayed harvest decreases the juice content and the keeping quality of the fruits (Ladaniya 2001).

A citrus fruit stays on the tree for 6 to 12 months; in the sub-tropics it may stay even longer (Singh and Naqvi 2001). Citrus fruits are mostly harvested manually using ladders. The existing practices of Kinnow harvesting in Punjab state are:

- Twisting, pulling and throwing on ground
- Twisting, pulling and putting in crates or in bags
- Cutting with secateurs and putting in crates

During 1990's the area under Kinnow crop in the state was almost double the area that was under the orange and malta cultivation put together. Kinnow production in Punjab during 2002-03 was 2,27,325 tonne (Anonymous 2003). The district wise production in tonne for the same period was Firozpur (107745), Hoshiarpur (46110), Muktsar (33765), Bathinda (13200), Rupnagar (7320), and Gurdaspur (390). The increasing trend in Kinnow growing is making it difficult for the farmers to sell whole of their produce in the local market. Therefore, the farmers are looking for distant markets such as Bangalore and Kolkata where they can earn more profit. There also exists export potential to Bangladesh, Kenya, Mauritius, Sri Lanka, UK and other countries to which India is exporting oranges at present (Vedini 2001).

In Punjab during peak harvesting period of Kinnow fruit, the growers do no get adequate price for their produce. The fruits can be stored conveniently in cold storage. However, such facilities are limited in India and beyond the reach of ordinary farmers (Reddy *et al.,* 1999). Under ambient storage conditions, fruits lose their turgidity in a short span of time and hence have shelf-life of 7-10 days even in winter months (Mahajan *et al.,* 2002, Randhawa *et al.,* 1999). This compels the growers to sell their produce at lower price in the local market. Under Indian conditions there is an urgent need to develop suitable techniques for consumer packing of mandarin fruits intended for fresh fruit markets (Ladaniya *et al.,* 1999).

The district wise production of Kinnow fruit and the location of waxing and grading facilities in Punjab are shown in Fig 1.1.

Wherever possible, the fruits should be pre-graded in the orchard; this reduces expenditure during pack-house operations. All bruised and damaged, too small, too big, hard or overripe and blemished fruits should be removed at this stage (Singh and Naqvi, 2001). Fresh horticultural produce undergo vigorous biological activities after harvest that accelerates loss of fruit tissues. Therefore, fruits tend to lose water at room temperature, change appearance, texture, and quality, and thus their commercial value decreases.

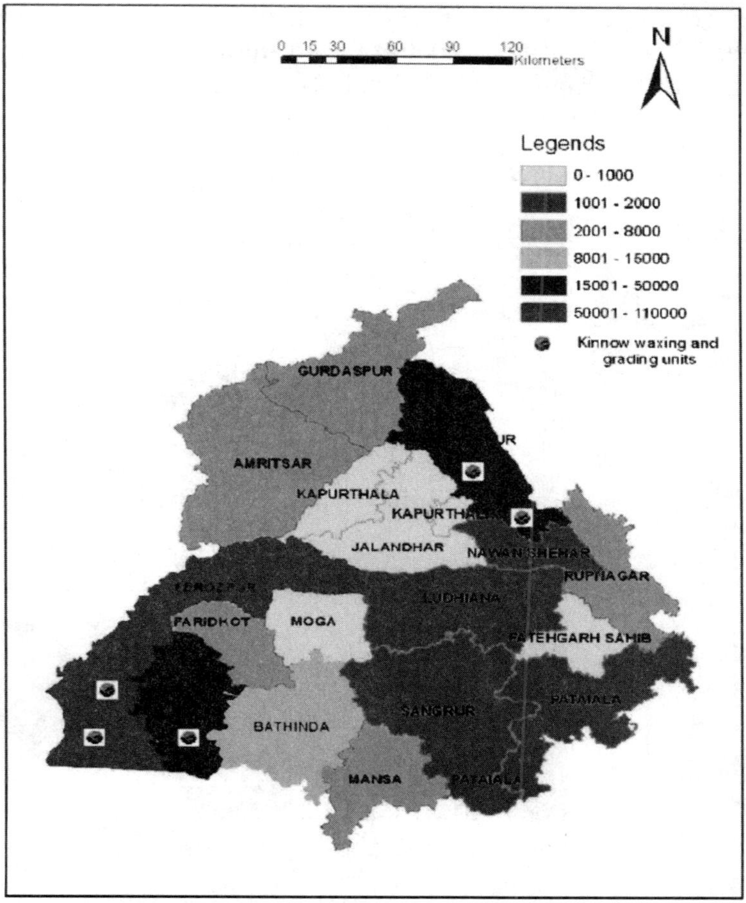

Fig 1.1. Production of Kinnow fruit (tonne) and the waxing and grading units in Punjab

STATUS OF POST-HARVEST PRACTICES FOR KINNOW IN PUNJAB

Farming practices

The orchard size pattern for Kinnow fruit in Punjab state is shown in Fig 1.2 The figure reveals that 18%

farmers have less than 5 acres of farm, 32% have 5 to 10 acres of farm and 50% farmers hold more than 10 acres of farm. The average orchard size is of 11 acres.

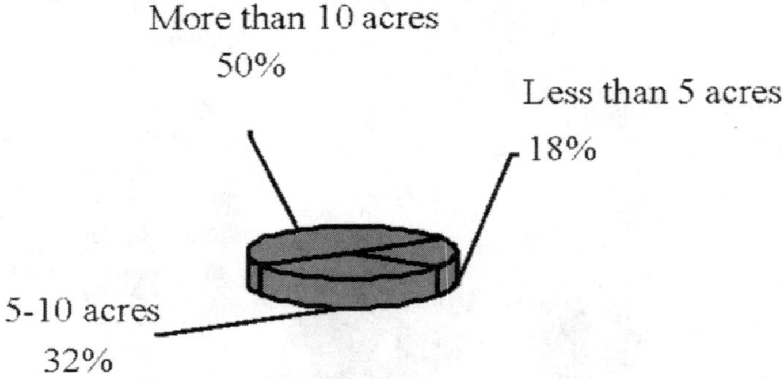

Fig 1.2 Classification of Kinnow orchards in Punjab by size

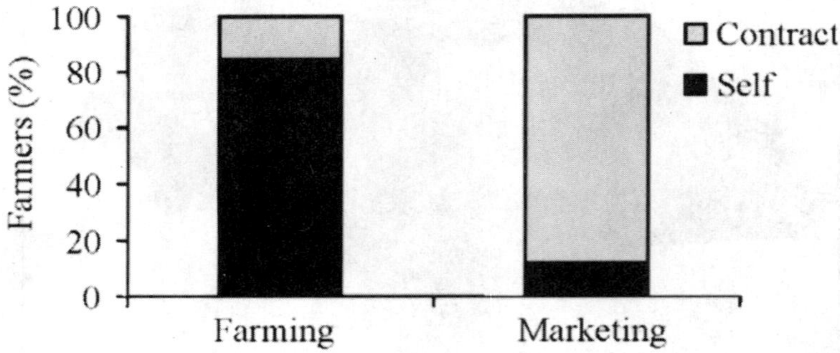

Fig 1.3 Farming and marketing trends of kinnow fruit in Punjab state

Kinnow growers of Punjab give their orchards on lease to the contractors. The Fig. 1.3 shows that nearly 90% of the Kinnow orchards remains on lease contract in one form or the other. The lease contracts are generally of two types. In the first type both growing and marketing of the produce is to be done by the contractor. The contract is signed for different time period. The period may range from one year to five years depending upon the size of the orchard. In the second type of contract only marketing of the produce is to be done by the contractor. In this type, the contract is signed for one season only and is generally signed prior to the harvesting season but after the setting of fruit. The second type of contract is more common than the first type. About 10% of the Kinnow growers carry out all the operations i.e. farming and marketing of Kinnow themselves.

The majority of the farmers bought the Kinnow nursery from local private nurseries and a small number of farmers buy it from State Government run nurseries. The planting pattern of nursery followed is 6.71 m x 6.71 m, 6.71 m x 6.1 m, 5.49 m x 6.1 m, 3.35 m x 3.96 m, and 3.35 m x 3.35 m. In the Abohar subdivision of Firozpur district 6.71m x 6.71m and 6.71m x 6.1m planting distance is followed whereas in the Hoshiarpur district there is no set pattern followed by the farmers. The farmers believe that narrow planting of nursery affects the life span of the orchard.

The status of power sources at orchards is that 71% farmers have tractors and 24% farmers have both tractor as well as electrical motors whereas 5% farmers have only electric motor as power source on the farm. The average distance of the orchards from the nearest motorable road is 10 km.

HARVESTING AND POST-HARVEST HANDLING

In Punjab, the harvesting of Kinnow fruit is invariably carried out by manual picking. The various methods followed are (i) twisting, pulling and throwing on ground, (ii) twisting, pulling and putting in crates, and (iii) cutting with secateurs and putting in crates. On an average it takes about one man-day to pick 625 kg of Kinnow fruit. In Firozpur district both local as well as migrant labour is employed for picking the fruit whereas in Hoshiarpur district only the local labour is employed. The female workers are paid about 30% less wages as compared to the male workers. The lower working efficiency is the main reason for the difference in wages.

The awareness of farmers regarding the washing, grading, waxing, and packaging of fruits is very low. The practice of wet-cleaning is not followed; and instead the individual fruit is wiped-clean using a piece of cloth. Nearly, 30-50% of the picking time is consumed in cleaning process; and nearly equal time to grade the fruits. Generally, contractors themselves do the grading job.

The practice of waxing of Kinnow fruit is followed only by those farmers/contactors who are engaged in sending their produce to the distant markets such as Kolkata and Bangalore. Such farmers are few in number as compared to the total number of growers. The interviews of some of the farmers at Chhauni Kalan (District Hoshiarpur) and Tahliwala Jatan (District Firozpur) waxing and grading unit revealed that the farmers have to arrange their own workers for loading and unloading of fruits at waxing units. Some times, during the peak harvesting season they have to wait for one to two days for their turn to come and it lead to increase in post-harvest handling cost of the fruits as they have to pay the workers for the waiting period also. Because of high initial cost, large land requirements to set up these units, short harvesting period, and due to non availability of proper storage facilities for the fruits, it is not feasible for the farmers to own such units on private basis. The Table 1.1 shows that the farmers with small orchard size of less than 5 acres are more inclined towards the use of waxing technology. This can be attributed to the fact that under this category 86% farmers are engaged in self-farming as well as self marketing of Kinnow fruit. The 100% contract marketing under the orchard size of 6 to 10 acres reveal that the contractors do not support the concept of waxing of Kinnow fruits for marketing. For the orchard size more than 10 acres, 44% farmers and contractors responded assertively for the need of waxing technology. The available information shows that the farmers seemed to be enthusiastic towards the adoption of on-farm waxing and grading technology if made available at affordable price.

MARKETING OF KINNOW

The status of marketing of Kinnow fruit shown in Table 1.1 reveals that in the farmer's category having orchard size less than 5 acres; 14% farmers follow the contract practice whereas 86% farmers indulge in self-marketing of their produce.

Table 1.1 Status of Kinnow farming and post harvest operations being followed in Punjab

Orchard size (acres)	Farming (per cent)		Marketing (per cent)		Waxing of fruit (per cent)	Post harvest operations followed	Need for waxing technology (per cent)
	Self	Contract	Self	Contract			
Less than 5	86	14	86	14	28	Cleaning, grading packing	71
6 -10	75	25	-	100	-	Cleaning, grading packing	-
More than 10	85	15	7	93	-	Cleaning, grading packing	44

The main reasons stated by the farmers for such pattern are (i) lack of knowledge about the market, (ii) less time available for sowing of other crops, (iii) financial constraints, (iv) cheating by the traders and (v) labour constraints.

Almost all the farmers having orchard size between 6 to 10 acres are following the contract marketing strategy. The farmers having orchard size more than 10 acres; about 93% are following the contract pattern whereas only 7% are engaged in self marketing of Kinnow fruit.

The delayed payments by the traders, cheating and the market fluctuations are the reasons stated by the farmers/ contractors behind the non-expansion of Kinnow market to the other states. Regarding market realization of waxed and graded fruit the farmers/contractors are satisfied as they earned a profit of approximately Rs. 1-2 fruit. According to the information provided by Punjab Agri Export Corporation Ltd., Chandigarh, India, during 2004-05 approximately 30 to 40 farmers/contractors are availing of the Kinnow waxing facility in five Kinnow waxing units situated in Punjab. The total quantity of Kinnow fruit waxed in these units during 2004-05 was 5500 tonne which is just 2% of the total Kinnow production during the same period (Fig. 1.4). Each unit caters to about 20 to 30 km of radius.

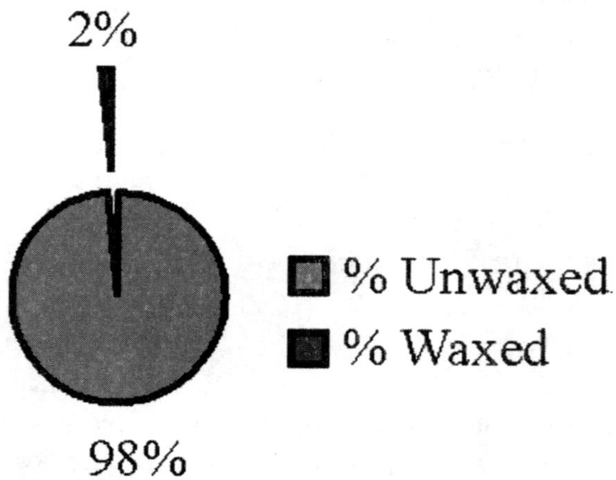

Fig. 1.4. Status of Kinnow fruit waxed in Punjab during 2004-05

For marketing fruits in the local markets, the fruits are packed in *Palli* (Jute sheets), wooden baskets, and sometimes packed in plastic crates. Wooden and corrugated fiberboard boxes are used only for distant markets. Tractor trolley is used as a transport means for local market whereas trucks are used for distant markets. It takes approximately 4 man-days to load a truck.

Chapter 2

ENHANCEMENT OF SHELF-LIFE OF FRUITS AND VEGETABLES

Fruits and vegetables are living entities. Their condition and marketable life are affected by temperature, humidity, the composition of the micro atmosphere which surrounds them, the level of damage that has been inflicted on them before, during and after harvest, and type and degree of infection with micro-organisms, insects, etc. (Thompson 1996). Where the considerable periods of storage are required to market the produce it is essential that the produce is of optimal physiological age and kept under the optimal environmental conditions (temperature, humidity and atmospheric conditions) to slowdown the post-harvest changes (ripening, senescence, water loss and pathogen growth) (Mitra 2005). Efforts have been made by various researchers to extend the storage life of Kinnow fruit (Mann 1978, Jawanda *et al.,* 1978, Sandhu *et al.,* 1983, Mann and Sandhanwalia 1977, Singh *et al.,* 1978, Lotha 1992, Nagar 1993, Sethi *et al.,* 1980, Bhullar *et al.,* 1985). Fresh fruits and vegetables start deteriorating immediately after the harvest. The quality losses occur in terms of appearance, texture, flavor and nutritive value of the commodity. Post-harvest shelf-life of fresh fruits and vegetables depends upon the factors such as handling methods, relative humidity and temperature during handling, packaging, transportation, storage, and the micro-atmosphere which surrounds them.

In India, fruits are handled several times for sorting in orchards and local markets before and after auction. The fruits are either packed in wooden boxes or transported loose. Such manual operations lead to number of handlings which cause injuries in fruits and increase contamination. The mechanical operations are efficient and require less time and space, and maintain good quality of produce. Most of the perishables being soft and fresh are prone to the structural damage called bruising, puncturing, pressure bruising and frictional bruising because of inadequate packaging during transportation. These damages lead to complete deterioration of the perishables within a short time period; hence the selection of product specific packaging material is essential for proper handling of perishables (Kitinoja and Gorny 1998).

The relative humidity in the post-harvest environment not only affects water loss causing wilting or shriveling, but also affects the activity of decay causing organisms. Water loss brought about by stomatal transpiration, especially leafy vegetables and immature fruits and vegetables depends upon the environment in which the fruits and vegetables would be stored. Water loss of three to six per cent is generally enough to cause a noticeable loss of quality and value.

During the period between harvest and consumption, temperature control has been found to be the most important factor in maintaining the product quality. Generally, for each hour of delay between harvest and cooling, one day of shelf-life is lost (Cantwell 1998). Generally adopted cooling methods are room cooling, forced air cooling, evaporative cooling, hydro cooling, and use of ice (Thompson 1996, Kitinoja and Gorny 1998, Mitra 2005).

Besides handling methods, relative humidity, and temperature, the other factor that is of significance is the gaseous composition of the micro-atmosphere. The studies have shown that the low concentration of O_2 and high concentration of CO_2 enhance the shelf-life of fruits and vegetables. This technique is known as modified atmosphere (MA) or an alternative term, controlled atmosphere (CA) is also used (Gorny 1997, Woodroof 1986). The modified atmosphere and controlled atmosphere technique differ only in the degree of control; CA is more exact (Kader 1980). Many plastic films are available for packaging, but relatively few have been used to wrap fresh produce, and fewer have gas permeabilities that make them suitable to be used for modified atmosphere packaging (MAP). The modified atmosphere conditions are created inside the packages by the commodity itself and/or by active modification (Zagory and Kader 1988). Commodity generated or passive MA is evolved as a consequence of the commodity's respiration (O'connor *et al.,* 1992). If a film of suitable permeability is selected, it is possible to develop an optimum equilibrium MA within the package (Exama *et al.,* 1993, Ballantyne *et al.,* 1988). On the other hand active modification involves creating a slight vacuum inside the package and replacing it with a desired mixture of gases, so as to establish desired equilibrium MA (Zagory and Kader 1988). The other techniques to enhance the shelf-life of fruits and vegetables are film coatings (Thomas *et al.,* 1964, Andres1984, Alam1998, Mahajan *et al.,* 2003) and, shrink wrapping (Ben-Yehoshua 1978, Ben-Yehoshua *et al.,* 1981, Purvis 1983, Dhatt *et al.,* 1995, Ladaniya *et al.,* 1998).

CONTROLLED ATMOSPHERE (CA) & MODIFIED ATMOSPHERE (MA) STORAGE/ PACKAGING

Many workers have studied the changes in gas concentration from the packaging of fruits and vegetables in polymeric films (Heing and Gilbert 1975). The permeability properties of several films by examining the respiration/

permeation dynamics of packaged produce have also been analyzed (Hayakawa *et al.,* 1975, Heing and Gilbert 1975, Mannapperuma and Singh 1987). The heat of respiration of fresh produce in controlled atmosphere storage was found to be 30% that of in air (Toledo *et al.,* 1969). The effect of modified atmosphere on respiration rate of daffodils was reported by Parson *et al.* (1967). Similar results under controlled atmosphere for Red Delicious apples were reported by Bohling and Hellickson (1998). The study indicated that reduction in the oxygen concentration from two per cent to one per cent caused a decrease in respiration rate of stored Red Delicious apples by approximately 30% while CO_2 and temperature were held constant at one per cent and 1°C, respectively. Increasing CO_2 level to two per cent, while maintaining the temperature at 1°C, reduced the respiration activity

Bhullar and Farnaham (1980) reported that an increase in the thickness of polyethylene bags was associated with a reduction in their permeability to moisture and air, thereby reducing physiological loss in weight (PLW) probably due to reduction of moisture loss and respiration of the produce. Polyethylene packaging maintained higher humidity inside the package which in turn, slowed down the process of evaporation and proper balance of CO_2 and O_2 concentrations, which further slowed down the respiration process (Hardenburg, 1971).

Geeson *et al.* (1985) revealed packs sealed with several polyvinylchloride (PVC) films or with K-resin (a butadiene - styrene copolymer) developed an atmosphere containing three to nine per cent CO_2 and three to nine per cent O_2 within 3-4 days at 10 or 12.5°C. Ripening of partly ripened tomatoes was retarded. In packs sealed with less permeable films (cellulose acetate) the internal atmosphere equilibrated to 10-18 per cent CO_2 and less than two per cent O_2 resulting in complete inhibition of ripening, a high incidence of rotting and tainting of fruit. Partially ripened tomatoes packed in permeable films gave 7 days longer life at ambient temperature than those without wrapping (Geeson, 1989).

Ben-Yehoshua (1985) developed individual seal packaging device for postharvest handling of fruits and vegetables for extension of self-life, reduction of shrinkage and weight loss. Pre-treatment of fruits and vegetables against decay was the prerequisite for MAP. Senescence of tomatoes was delayed to one week at 20°C with ethylene absorbent in this technique. The high density polyethylene (HDPE) films reduced water loss by 13.8 times without inhibiting gas exchange.

For many fruits and vegetables, the respiration rate and the moisture build-up rate doubled with a 5°C rise in temperature (Harderburg *et al.,* 1986), but this rise would increase film water permeability by 23 per cent. This implies that although a bag may adequately increase its oxygen transmission rate (OTR) to keep a product aerobic for a given storage temperature rise, it may not adequately increase its water permeability to prevent free moisture build up in the bag, which would encourage spoilage (Zagory and Kader, 1988).

Kubo *et al.* (1990) concluded that in non-climacteric fruits (3 citrus species, grapes and Japanese pears) and several vegetables (carrots, onion, cauliflower and cabbage) in which C_2H_4 production was not detected, high CO_2 affected respiration little if at all. Fruits and vegetables which when placed in high CO_2 atmosphere and produced C_2H_4 showed an increased respiration while high CO_2 decreased respiration only in climacteric fruits and broccoli, which were producing C_2H_4. They indicated that the respiratory responses of harvested horticultural crops to high CO_2 might be mediated by the effect of CO_2 on the action and/or synthesis of C_2H_4.

Two types of MA packages namely; rigid and flexible are commonly used to market horticultural commodities.

Rigid packages are made with an impermeable rigid material. Atmosphere modification in the package is achieved by heat sealing a semi impermeable window made of a polymeric film on the top of the rigid shell. Flexible packages are totally made of semi permeable polymeric films which are heat sealed after filling with the produce. For soft and more susceptible to injuries products, rigid packages are used, whereas flexible packages are more popular for marketing pre cut vegetables and salad mixes in developed countries. The design of a MA package involves specification of the gas permeabilities, thickness and area of the required packaging film, mass of the product to be placed in the package and dimensions of the package for any given surrounding temperature and optimum gas concentration. Time required for the packaging system to attain the optimum conditions is also an important consideration in designing MA packages, especially for products whose storage life is short (Talasila *et al.,* 1992). If the time required to attain the optimum conditions is long relative to the storage life, the packaged product may not get the full benefit of MA packaging.

Exama *et al.* (1993) concluded that all the films low density polyethylene (LDPE) and PVC could satisfy the O_2 fluxes for low respiration rate produce such as apple, carrot, celery, cabbage and green pepper with only minor changes in film thickness of area to weight ratio. For produce with moderate respiration (cauliflower), most of the polymers would not be adequate. For produce with higher and very high respiration (mushroom), the membrane developed by Marcellin (1974) would be the only polymer among those examined which could provide the required high fluxes. The required gas selectivity (S^R) for the majority of films was much higher (>6) while most produce required S^R between 1 and 6. Only combination of polymeric and perforated films could potentially provide fluxes and selectivity for commodities with high respiration rate. These workers further suggested an alternate solution, a device to act as a safety valve. It helps in preventing excessive O_2 depletion and excessive CO_2 accumulation when a transient temperature increase occurred.

Moleyar and Narsimham (1994) emphasized the importance of MAP in the present era of consumer awareness of residues of pesticides, additives and preservatives in food and of economic need to extend the storage life of perishables. They suggested a modified atmosphere of O_2 about two per cent and CO_2 level 3.5% for the storage of fruits and vegetables in polymeric films of different gas permeabilities. The Kinnow fruits under CA storage with elevated percentage of CO_2 gave better results at 5°C. Under CA storage with reduced level of O_2, produced alcoholic taste after 100 days of storage. For storage of Kinnow fruits, three per cent level of CO_2 at 5°C was found to be the most ideal combination (Singh 1990). The broccoli florets washed with 200 ppm NaOCl and held at 0,5,10, and 15°C in several sealed Cryovac films were capable of producing MA compared with controls in perforated polyethylene film bags. The Cryovac PD-941 was found to be the most effective for retaining excellent quality for at least 4 weeks at 0°C. In some low permeable films severe off-flavors were apparent especially at higher temperatures (Cabezas, 1997). To overcome the problems of off-flavour development under MAP, films with higher permeability, more sensitive to change in temperature, in which the relative permeability of CO_2 to O_2 can be altered i.e. membrane technology were reported by Clarke *et al.* (1997). The increase in permeability reduces the chances of micro-atmosphere inside the package becoming anaerobic during the incidences of temperature change when the cold chain can not be maintained at the retailer end. Anonymous (1992) produced first controlled respiration and ethylene absorption polyethylene film (Fresha - Pac). This type of film is produced with different permeation and absorption rates to suit a wide range of vegetables, fruits and cut flowers for shipping and arriving fresh in the world.

Christie *et al.* (1995) determined the film permeability by using the material balance equations for the conditions in which the produce is to be stored. They also measured O_2 consumption, CO_2 and ethylene production at 8.8 ml/kg/h, 9.0 ml/kg/h and 0.04 μl/kg/h respectively for broccoli stored in MAP (6% CO_2, 1.5% O_2) at 1.5°C. They recommended mixed gas cell method to have realistic prediction of film permeabilities. Also, Akimoto and Maezava (1997) reported a method to measure respiration rate in which both the respiration of the produce as well as the permeability of the fruit were considered. This could be used over the whole storage period including steady and unsteady states. This method also predicted the respiration quotient for perishables in MAP under unsteady state conditions.

The nine types of polyethylene films for modified atmosphere packaging (MAP) for O_2, CO_2 and water vapour transmission rates at four temperatures (0.0, 5.0, 15.5. and 23.2°C) as evaluated by Moyls *et al.* (1998) have shown that the average ratio of CO_2TR/OTR decreased from 5.93 to 5.26 as temperature increased from 0 to 23.2°C while the average ratio of H_2O/O_2 permeability (on molar bases) decreased from 80.7 to 47. Moyls (1998) described whole bag (water) method as more accurate, flexible, fast, easy and comparatively less expensive method for water vapour transmission rate (WVTR) for the entire bag surface rather than for a small piece. The WVTR can be determined at any desired temperature, however requires a supply of air at constant relative humidity (Rh) and temperature for more reliable results for polyethylene and other hydrophobic films.

Sudheer *et al.* (1998) used the silicone membrane technology for a possible alternate for the controlled atmosphere storage. MA conditions were maintained membrane installed as windows on laboratory scale containers. After

60 days, MA stored mandarins had better freshness, appearance and lower weight loss (1-2%) than those stored for the same period under regular atmosphere (weight loss of 30%), The MA stored strawberries also showed less physiological loss in weight (1.4%) and a firmness reduction of 34% at the end of 14 days storage using silicone membrane.

INDIVIDUAL SHRINK WRAPPING OF FRUITS AND VEGETABLES

The Kinnow fruit need proper storage so as to extend shelf-life for longer duration for consumption. Pre-packing of the fresh fruits in plastic bags is also helpful in extending the shelf-life both at room temperature and cold storage. Jawanda *et al.* (1978) reported that the loss in weight of the whole fruit, peel and pulp was markedly reduced by storing in polyethylene bags. The weight loss was 6.3 to 8.7 and 48.4% respectively with and without pack. The TSS and the reducing sugars increased while acidity declined slightly upon prolongation of storage in poly bags. Such fruits also retained firmness and usual taste, and flavour during storage. Postharvest life of the Kinnow fruit can be prolonged to nearly 90 days when stored in perforated polyethylene bags at a temperature of 4 to 4.5°C and Rh of 85 to 96%. Singh (1981) also suggested similar low temperature storage of 2.2 to 3.0°C for polyethylene packed Kinnow fruits to keep the fruits satisfactorily over a period of 60 days. Seal-packaging of individual citrus fruits with shrinkable polyethylene film, successfully doubling the shelf-life was also reported by Ben-Yehousha (1978). Since then several researchers reported the usefulness of shrink wrapping in citrus fruits.

Reduction in moisture loss from oranges and grapefruit stored at 21 and 5°C was reported by Purvis (1983). For green capsicums wrapped in plastic films it was reported by Hughes *et al.* (1981) that the produce remained fresh for longer, and lose less weight than other treatments under study. The weight loss was about three times greater in hypobaric stored capsicums. The fruits became wrinkled and flaccid due to severe desiccation. Storage temperature of 4°C with high humidity in sealed polyethylene sleeves, for maximum of 4 week storage for gladiolus spikes without any reduction in vase life was reported by Nowak and Rudnicki (1984). Plastic film performed better than the waxing and anti transpirants in enhancing the shelf-life of the Valencia and Navel oranges (Jimenez *et al.,* 1983). Shrink wrapping of grapefruit in plastic films resulted in reduced weight loss, retained firmness under non-refrigerated conditions and good quality at the consumer end was reported by Langston and Pfuntner (1986).

The use perforated polyethylene bags of 100 gauge thickness to pack Kinnow fruits for storage at 0-3.3°C and 85-90% Rh by Bhullar *et al.* (1985) showed rotting of fruit after 85 days of storage. At higher storage temperature of 17-22°C and 60-65% Rh, polyethylene wrapped Kinnow fruits showed 69.6% spoilage as reported by Sandhu *et al.* (1983). Minimum spoilage 13.6% was found in fruits treated with 30 ppm of GA. Prolonged storage of Kinnow fruits at 3.3°C by Farooqui *et al.* (1981) showed spoilage due to infection by *Penicillium citricum*. Ben-Yehoshua *et al.* (1981) reported that seal packaging of individual fruit resulted in much less decay than sealing a whole carton of fruit together. Also, the fungicides imazalil, sec-butyalamine 2-phenylphenol benomyl and thiabendazole markedly reduced the decay of sealed fruit in all cultivar of citrus fruits tested.

Shrink wrapping of fresh sweet corn stored under refrigerated conditions resulted in 3-fold extension in shelf-life (Deak *et al.,* 1987). The fungicide treated mature green and pink tomatoes were individually shrink wrapped with plastics films and stored at 13 and 18°C with unwrapped (control) tomatoes. Pink tomatoes lost less water as compared to unwrapped and the shelf-life of mature green tomatoes increased by 10 days by shrink wrapping (Hulbert and Bhowmik 1987). Similarly, washed, fungicide treated, mature green tomatoes (cultivar Sunny), individually shrink wrapped with single or double layer of a heat-shrinkable plastic film (60 gauge Cryovac D-950 film) stored at 13°C and 85-90% Rh with the unwrapped (control) tomatoes were found better than the control and last for 60 days (Bhowmik and Hulbert, 1989). The investigations regarding the shelf-life of individually shrink wrapped peaches in heat shrinkable polyolefin film (60 gauge) was determined as 5-6 week at 1 and 10°C, as compared to 3-4 week at 1°C for control peaches (Bhowmik and Sebris, 1988).

Heaton *et al.* (1990) monitored the changes in quality of unwrapped and wrapped cultivar Starkrimson apples during a 38 week storage period (26°C and 40-42% Rh). Shrink wrapped apples were subjectively judged to have acceptable qualities throughout storage.

Kinnow fruits packed in 0.010 mm thick HDPE film bags and stored at ambient temperature (7-32°C) maintained their appearance and were rated tasty after 8 weeks of storage as compared to unwrapped fruits which were judged non-edible due to development of off-flavor (Dhatt *et al.,* 1991). Kinnow fruits dipped in 1000 ppm imazalil solution and wrapped in 10 micron thick HDPE bags can be stored successfully to 60 days at ambient temperature without much loss in fruit weight and fruit quality (Dhatt *et al.,* 1995). Wrapping of 'Nagpur mandarin' fruits in heat-shrinkable polyethylene (25 and 11 μm), BDF 2001 (30 μm) and D955 (25 and 11μm) films after chlorine (1000 ppm) wash and bavisitin (4000 ppm) treatment extended the shelf-life up to 3 weeks at ambient condition and reduced the weight loss in refrigerated storage without losing their natural appearance and flavour (Ladaniya *et al.,* 1998).

Eureka lemons (*Citrus limon* (L.) Burm. F.) sealed HDPE plastics film showed overall marketing quality higher than the unsealed lemons 6 months and further recommended the installations of seal packaging lines for extended storage of lemons (Cohen *et al.,* 1990).

APPLICATION OF FILM COATINGS

Materials available for forming films and film coatings fall generally into the categories of proteins, polysaccharides, lipids and resins. A plasticizer must often be added to reduce film or coating brittleness. A surface-active agent is also often necessary to aid film or coating formation. Other constituents can include antioxidants and antimicrobials to enhance the film or coating effectiveness. The U.S. Code of Federal Regulations provides the status of protein, polysaccharide, lipid, resin, plasticizer, Emulsifier, preservative and anti-oxidant materials related to acceptable use (Baldwin, 1999). The three types of waxes tested and commercially used for fruits and vegetables such as natural waxes-beeswax, lanolin, shellac wax; vegetable waxes-carnauba, candelilla, bayberry, resin wood, cottonseed oil, sperm oil, palm oil, and rice bran oil; and mineral waxes-molten paraffin, and microcrystalline wax were listed (Mahajan *et al.,* 2003). The Kinnows coated with protein based coating – casein (8%) and the cellulose based coating – carboxy methyl cellulose (0.5, 1.0, and 1.5%), and packed in high density polyethylene bags revealed minimum physiological weight loss, maximum firmness and juice percentage in fruits coated with 0.5 per cent CMC and packed in high density polyethylene bags (Alam, 1998). Similar results were quoted by Sharma (1999) in his study on effect of edible coatings on shelf-life and quality of okra. Lerdthanangkul and Krochta (1996) studied the green bell pepper fruits coated with a mineral oil based coatings and a cellulose based coatings. The effects of these coatings were followed by measurement of changes in respiration, internal gases, colour, firmness, and water loss during storage at 10°C, 80-85% Rh for 20 days. Only the mineral oil based coating significantly reduced the moisture loss, thus maintaining fruit firmness and thereby prolonging fruit freshness.

Proteins cover a broad range of polymeric compounds that provide structure or biological activity in plants or animals. Proteins arc distinguished from polysaccharides because they are based on approximately 20 amino acid monomers rather than just a few or even one monomer, such as glucose in the case of cellulose and starch. The amino acids are similar in containing an amino group (-NH_2) and a carboxyl group (-eaOH) attached to a 'central carbon atom. However, each amino acid has a different side group attached to the central carbon that lerids unique character to that amino acid. The side group can be non-polar (hydrophobic), polar uncharged (hydrophilic), positively charged at pH 7, or negatively charged at pH 7 (Cheftel *et al.,* 1985).

Most proteins contain 100-500 amino acid residues. Depending on the sequential order of the amino acids (primary structure of the protein), the protein will assume different structures along the polymer chain (secondary structure of the protein), based on Van der Waals, hydrogen bonding electrostatic, hydrophobic, and disulfide cross-link interactions among the amino acid units (Cheftel *et al.,* 1985). The tertiary protein structure reflects how the secondary structures organize relative to each other, based on the same types of interactions, to form overall globular, fibrous, or random protein structure. Finally, quaternary structure occurs when whole proteins interact with each other into associations to provide unique structure or biological activity.

The secondary, tertiary, and quaternary structures of proteins can be modified by various physical and chemical agents, including heat, mechanical treatment, pressure, irradiation, lipid interfaces, acids and alkalis and metal ions (Cheftel *et al.,* 1985). Such agents are often used in the formation of protein films and coatings to optimize protein configuration, protein interactions, and resulting film properties. Protein film-forming materials derived from animal sources include collagen, gelatin fish, myofihrillar protein, keratin, egg white protein, casein, and whey protein. Protein film-forming materials derived from plant sources include corn zein, wheat gluten, soy protein, peanut protein, and cottonseed protein.

Polysaccharide film-forming materials include starch and starch derivatives cellulose derivatives, alginate carrageenan Chitosan, pectinate and various gums. Proteins can be combined with polysaccharides to modify film mechanical properties (Shih, 1994; Arvanitoyannis *et al.,* 1996, 1997, 1998a, 1998b; Arvanitoyannis and Biliaderis, 1998).

Edible lipids include beeswax, candelilla wax, carnauba wax, triglycerides (e.g. milkfat fractions), acetylated mono glycerides, fatty acids, fatty alcohols, and sucrose fatty acid esters. Edible resins include shellac and terpene resin. Because lipid and resin materials are not polymers, they do not generally form cohesive stand-alone films. However, along with often providing desirable gloss they can be used to coat a surface to provide a moisture barrier or to provide the moisture-barrier component of a composite film. Composite films can consist of a lipid layer supported by a protein or polysaccharide layer, or lipid material dispersed in a protein or polysaccharide matrix (Krochta, 1997a).

Protein films and coatings are often quite still and brittle due to extensive interactions between protein chains through hydrogen bonding, electrostatic forces, hydrophobic bonding and/or disulfide cross-linking. Relatively small molecular weight hydrophilic plasticizers are often added which mainly compete for hydrogen bonding and electrostatic interactions with the protein chains. The result of plasticizer addition is a reduction in protein chain-lo-chain interactions, a lowering of the protein glass transition temperature(s) and an improvement in film flexibility (lowering of film elastic modulus), Also, film elongation (stretchiness or ductility) increases, and film strength decreases. Unfortunately, plasticizers generally also decrease the film's ability to act as a barrier to moisture, oxygen, aroma, and oils, Plasticizers acceptable and generally used for protein edible films include glycerol, propylene glycol, sorbitol, sucrose, polyethylene glycol, fatty acids, and mono glycerides. Water is also an important plasticizer for protein films. Thus, film moisture content, as affected by the surrounding environment's relative humidity (RH), has a large effect on film properties. The presence of hydrophilic plasticizers such as glycerol at extracts additional moisture and additionally impacts film properties, Table 2.1 shows plasticizers that are commonly used with protein films and coatings.

Emulsifiers are surface-active compounds with both polar and non-polar character, which absorb at the water-lipid interface and reduce surface tension. To produce protein-lipid or polysaccharide-lipid composite films from aqueous solution, it is often necessary to add an emulsifier to allow dispersion of the lipid material in the solution. Also, for food-coating applications, addition of a surface active agent to a coating formulation ally is necessary to achieve surface wetting and spreading with the coaling formulation and then adhesion of the dry coating. Some protein, are sufficiently surface active that no emulsifier is necessary to form well-dispersed composite films or provide good surface wetting and adhesion.

As mentioned, films and coatings have the potential to be selective carrier, and provider, of antioxidant, anti-microbials, nutrients, flavor and colors to enhance food safety, nutrition and quality.

A polysaccharide material, such as starch can be combined with protein to produce a blended film system. Limited data exist on such blended films (Arvanitoyannis and Biliaderis, 1998).

The greatest reduction in protein film WVP is achieved by addition of a lipid or wax to the film structure. The protein provides structural integrity for the film while the hydrophobic lipid or wax provides the moisture barrier Dispersing a lipid or wax in the protein film structure to achieve an "emulsion film" can reduce the film WVP by an order of magnitude. However, this still places the film's WVP one to three orders of magnitude greater than the WVP of the pure lipid or wax, or of a polyethylene film. Thus far, the only way to achieve a good moisture barrier is to laminate a pre-formed protein or polysaccharide film with a layer of lipid or wax to achieve a bi-layer film. While addition of a lipid or wax to a protein film can reduce film WVP, such addition usually increases film 0P, reduces film TS and EM and increases film E. Also, the possibility of resulting cracks and pinholes in the film increases.

Waxing of Kinnow fruits in combination with benlate 50 was found effective in minimizing the physiological weight losses. Dhillon *et al.* (1977) applied wax plus benlate 50 to improve the cold storage life of Kinnow. Waxol (12%) treatment was reported as the best by Singh *et al.* (1973) for retaining highest quality of Kinnow fruit for 35 days. Similar investigations by Farooqui *et al.* (1988) on the effect of two commercial wax emulsions viz. Britex-561 and SB-65 (USA) on the storage behaviour of Kinnow fruit revealed that waxing not only improved the external appearance of the fruit but also reduced weight loss, slowed down the respiratory activity and ethylene production, thus extending life. Earlier studies by Thomas *et al.* (1962) on yellow limes treated with aqueous emulsions containing (A) carnauba wax, sugarcane wax and bleached lac and (B) emusifiable polyethylene; revealed that fruits treated with emulsion (A) maintained good quality for longer period under ambient conditions (25-33°C, 70-80% Rh). The coating of fruits with wax emulsions containing solids more than 8% resulted in physiological breakdown. Hagenmaier (2001) studied the mandarin hybrids treated with wax and resin coatings having differing oxygen permeability. After storage for 7 days at 21°C, fruit with coatings having O_2 permeability $\leq 1.1 \times 10^{-16}$ mol m s^{-1} m^{-2} Pa^{-1} were rated by a sensory panel as markedly less fresh than fruit with higher permeability coatings. These low-permeability coatings were those composed mainly of shellac and wood resin. The flavor changed most for fruit having mean internal $O_2 < 4\%$, internal $CO_2 > 14\%$ at 21°C and juice ethanol content >1500 ppm after 7 days storage at that temperature. All coatings seemed suitable for storage of specialty citrus fruit for 7 days at 5°C. It is recommended that for mandarin hybrids, the most suitable are those composed mainly of waxes rather than shellac or wood resin.

Miller and Grierson (1983) evaluated the properties of preservative coatings on citrus fruits. The properties studied were pH, solids, specific gravity, specific heat, water vapour transmission, and viscosity. The pH was measured using Corning Model 12 pH meter. Solids (per cent) were determined by drying the samples at 105°C for 24 hr. The specific gravity was measured by conducting tests with specific gravity/Baume hydrometer at 15 and 25°C. For specific heat the law of mixture was used to determine the specific heat of 1600 ml coating samples. For WVTR the procedure based on desiccant moisture uptake similar to that described by ASTM (1980) was employed. The cross-section samples of the peel of Valencia oranges were observed under a light microscope to determine the thickness of the coating. A Brookfield viscometer with UL adapters was used to measure the coating viscosities at 15 and 25°C. The average thickness of natural wax was 1.48

μm, whereas the applied coating thickness was 6.34 μm for single application and 8.07 μm in a double application.

Andres (1984) treated fruits with protein (corn zein) based coating and vegetable oils. The coating was applied by dipping, misting, and spraying methods. The study revealed that these had good moisture barrier properties. Similar study by Park *et al.* (1994) on tomatoes revealed the extension of shelf-life by 6 days under 21°C storage conditions. Various fruits and vegetables treated with an edible coating showed the retardation in ripping process in tomatoes, mangoes, pumpkins, and bananas (Nisperos *et al.,* 1991). Carrots coated with edible coatings and stored at 2°C showed reduced level of surface discolouration and retained better flavour and overall acceptability than the non coated carrots (Howard and Dewi, 1995).

The study on the gas permeability of O_2, CO_2, and C_2H_4 and water vapor for 19 commercial fruit wax coatings by Hagenmaier and Shaw (1992) showed that a typical coating was of 0.008 mm (0.3 mil) thick. The permeability study was based on the theory of Fick's law. It was found that the commercial waxes had sufficiently low non-condensable gas permeability to account for large reductions in the respiration rate of the coated fruit. Wax coatings could be improved if permeability were controlled. Later on Hagenmaier and Baker (1996) reported an improved candelilla wax coating formulation with good gloss and satisfactory gas barrier properties. The wax was obtained from the desert plant *Euphorbia antisyphilitica,* which grows in the southwestern United States and northern Mexico. Water vapor permeability of candelilla wax was reported to be 55 per cent that of carnauba wax and 31 per cent that of beeswax (Donhowe and Fennema, 1993). Citrus fruits coated with candelilla micro-emulsions had low weight loss and were less glossy (Hagenmaier and Baker, 1994; 1995).

Veraverbeke *et al.* (2001) conducted the non destructive analysis of the wax layer of apple. Confocal laser scanning microscopy (CLSM) was used to analyze the changes in the structure and thickness of the cuticle during CA storage of apples. Interpretations of the confocal images were performed by comparison with SEM and environmental SEM images. Natural reflectance of the wax and auto-fluorescences of the under lying cells made it possible with CLSM to distinguish the wax from the under lying layers without any pretreatment of fruits. The thickness of consecutive layers (wax, cutin, cells) could be estimated from measurement of reflection and fluorescences intensities as a function of the number of pixels. The mean wax-layer thickness measured in this way amounted to 2.58, 3.41, and 4.14 μm for cultivar Jonagold, Jongored and Elstar, respectively.

Mc Guire (2002) evaluated *Candida guilliermondii* for compatibility with commercial citrus coatings for application to heat-treated grapefruits (*Citrus paradisi* Macf.) and biological control of *Penicillium digitatum*. Although constituents of many coatings were fungicidal, FMC 705 and FMC 214 (FMC Corp., Lakeland, FL) and Nature Seal (USDA product) were satisfactory carriers for the yeast. *Candida guilliermondii* survived for 2 months at 13°C within films of dried coatings on filter paper disks. When applied to grapefruits in FMC 705 or Nature Seal at concentrations calculated to produce initial surface populations of 10^4 or 10^6 cfu/cm^2, the populations stabilized over 9 weeks at 13°C between 10^5 and 10^6cfu/cm^2. Populations of unidentified, native *Candida* species on fruits were originally 300 cfu/cm^2 before a 3-h quarantine treatment in air at 48°C reduced them to 20 cfu/cm^2. Although these low native populations on heated fruits subsequently rebounded over 9 weeks to levels approximately 10^4 cfu/ cm^2, fruits to which *C. guilliermondii* was applied remained free of decay 12 to 20% longer in cold storage than fruits not treated with this species.

Raymond *et al.* (2002) reported that Shellac coatings which support significantly high populations of the yeast *Candida oleophila* for biocontrol of *Penicillium digitatum* were formulated for postharvest application to grapefruits. An aqueous shellac latex was not toxic to the yeast and was suitable for testing constituents of commercial shellac coatings. Alteration of this product to contain two per cent ethanol was nontoxic, but 4, 6, 8, and 10% ethanol killed 90% of the yeast cells within 24, 4, 2, and 0.5 h, respectively. Above a pH of 8.25, cell death was rapid when amine bases were incorporated. This level of toxicity resulted from addition to the shellac latex of more than 1.5 per cent morpholine or 0.3% ammonia; a similar death rate developed nearer to pH 9.0 with incorporation of approximately 1.5% KOH. Although higher concentrations of base were required for the preparation of experimental formulations derived from bleached shellac, shellac ester, and wood resin ester, the relationship between the pH of the liquid coating and yeast survival persisted. Particularly, with wood resin ester, KOH solutions maintained higher populations of yeast. When applied to grape fruits post-harvest, a shellac latex formulation and a dissolved shellac /shellac ester formulation supported surface populations of yeasts between 10^4 and 10^5 cfu/cm^2 throughout 4 months of storage at 13°C. These preparations significantly improved shelf-life when compared to fruits coated with a commercial shellac formulation.

Chien *et al.* (2006) reported the effects of coating with low molecular weight chitosan (LMWC, Mw = 15 kDa) and high molecular weight chitosan (HMWC, Mw = 357 kDa) on the decay of Murcott tangor and the maintenance of its quality. A 0.1% LMWC coating substantially slowed the decay of Murcott tangor stored at 15°C in relation to a control sample and reduced decay by over 20% as compared to the fungicide thiabendezol (TBZ). A concentration of 0.2% LMWC was more effective in controlling the growth of fungus on citrus fruits caused by *Penicillium digitatum* and *Penicillium italicum*, exhibiting effective antifungal activity. LMWC coating improved firmness, titratable acidity, ascorbic acidity and the water content for Murcott tangor stored at 15°C for 56 days. Consequently, Murcott tangor coated with LMWC exhibited greater antifungal resistance than TBZ and its quality was maintained for longer

The materials, properties and potential applications for edible films and coatings were reported by many researchers in the past (Daniels, 1973; Guilbert, 1986; Kester and Fennema, 1986; Krochta, 1992; Lindstrom *et al.,* 1992; Conca and Yang 1993; Cuq *et al.,* 1994; Koelsch. 1994; Krochta *et al.,* 1994; Baldwin *et al.,* 1995; Callegarin *et al.,* 1997: Gennadios *et al.,* 1997; Myllarinen et al, 1997; Debeaufort *et al.,* 1998: Baldwin, 1999). Both edible and biodegradable films and coatings were also reported (Ahvenainen *et al.,* 1997: Guilbert *et al.,* 1997: Krochta and De Mulder-Johnston, 1997: Arvanitoyannis and Gorris. 1999). Several reviews focused exclusively on either proteins (Gennadios and Weller, 1990; McHugh and Krochta, 1994; Torres, 1994; Krochta, 1997), polysaccharides (Nisperos-Carriedo, 1994; Nussinovitch, 1998), or lipids and resins (Hernandez, 1994). Films are considered to be stand-alone, being formed separate of any eventual intended use. These are used as testing structures for determination of barrier, mechanical, solubility, and other properties provided by a certain film material. Such films can be used as covers, wraps, or separation layers: and they can be potentially formed into casings, capsules, pouches, and bags. Related products include molded items of greater thickness. Coatings involve formation of films directly on the surface of the object they are intended to protect or enhance in some manner. Coatings become part of the product and remain on the product through use and consumption.

Edible wax and shellac coatings are used on fruits and vegetables to prevent moisture loss and improve appearance. Wax, shellac, and zein coatings are used on confections to provide moisture resistance and improve appearance. Sucrose esters of fatty acids are used in formulations to reduce respiration, moisture loss, and skin marking and bruising in fruits and vegetables.

Hydroxypropyl methylcellulose (HPMC), hydroxypropyl cellulose (HPC), methylcellulose (MC), shellac and zein coatings and gelatin capsules are used for pharmaceuticals to improve product appearance, structural integrity, and ingestibility, as well as to protect against moisture and oxygen. HPMC and MC minimize cost of coating materials and coating process. Reduce packaging needs are also used for coatings on fried foods to reduce moisture loss and fat gain. MC is also used in commercial formulations to extend the life of fresh whole or cut fruits and vegetables. HPMC pouches are marketed for delivery of pre-measured food ingredients to food formulations with subsequent dissolution of the pouch.

Starch-based materials are marketed for coating drug tablets. Starch and dextrin formulations are also marketed as protective coatings, integrity maintainers, appearance enhancers, and seasoning adhesives for nuts, snacks, cereals, and meats. Collagen casings and wraps inhibit moisture loss and oxygen transport and provide structural integrity for meat products.

Every produce suffers from at least one mass transfer problem, whether it be moisture migration, oxygen intrusion, aromaloss, or oil migration Many produces could also benefit from improvement in integrity or appearance. However, applications have not become more widespread because commercial film and coating materials are either too expensive or too difficult to form, and/or commercial materials do not function efficiently. Thus, extensive research continues in this area. Certain new edible film and coating materials, including proteins, are targeted for replacing materials currently used in existing applications. Beyond existing applications, edible films and coatings generally seen as having considerable potentia1 for applications far beyond current usage for improving product quality and, in some instances, reducing us of synthetic packaging films. Several reviews list edible protein film and coating applications that have been explored over the years (Baker *et al.,* 1994; Baldwin, 1994; Krochta, 1997; Krochla and Dc Mulder-Johnston, 1997). The information is generally lacking on approaches to coating foods, as well as the resulting effectiveness of edible films and coatings in food systems. This makes it very difficult for food processors to decide on the "value-added" merit of an edible film or coating relative to the additional cost involved.

Chapter 3

MECHANICAL HANDLING OF FRUITS AND VEGETABLES

Most of the equipment owners and suppliers have their own ideas about what they want to accomplish. The purpose is to enunciate basic principles that apply to mechanical handling of fruits and vegetables, regardless of the particular layout. Errors in selecting and matching packing line components are extra ordinarily costly in money, reduced capacity, and increased fruit losses and decay claims. Various researchers have outlined the basic packinghouse line, requirements, limitations, and the points for the improvement. The various operations underlined by the researchers are: dumping; cleaning-wet or dry; surface drying; color addition; waxing; and size grading. But, there is no fixed arrangement of these operations. The various pack-house operations are shown in Fig 3.1.

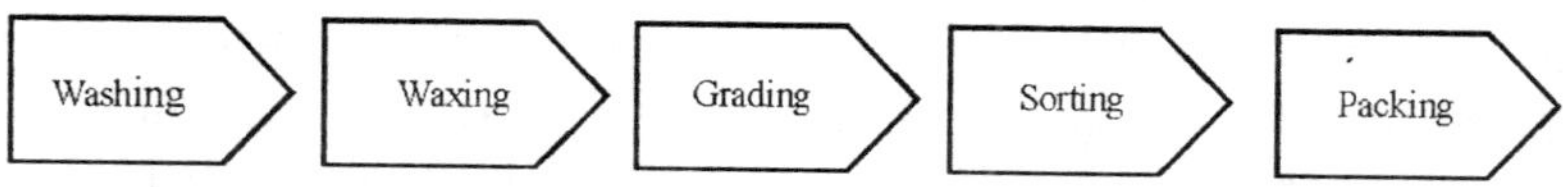

Fig. 3.1 Pack-house operations for fruits and vegetables

MECHANICAL HANDLING OF FRUITS AND VEGETABLES

Sharma and Kapur (1968) developed a hand operated village type waxing unit. The unit was consisted of a stainless tank, oval shaped at the bottom. The wax emulsion was filled in this tank. The capacity of the tank was 57 litre. A stainless steel basket was fitted into the tank. The basket can be raised or lowered down by a handle fitted with lever mechanism. The tank was placed on a stainless steel frame. At the same place where the basket was hinged to the tank, a cone type attachment with a stop gap arrangement for the delivery of fruits to the conveyor belt was provided. A groove has been made at the tapering end of the cone which ends in a hole of 0.7 cm diameter. This drains out the excess wax emulsion adhering to the fruits. Two chambers separated by a conveyor belt constituted the drying unit. The upper chamber has shutters at both ends. On top of it two holes were provided for noting down the temperature at both the ends. The lower chamber brings the hot air from the blower to the upper chamber. A conveyor belt, operated by a crank runs in between the two chambers. A blower which is fixed at the end of the lower chamber provides air flow to the fruits. The air was drawn in through an opening near the blower, under which a stove was placed to provide the heat.

A low cost waxing unit was also reported by Garcia and Galvan (2003). The machine was developed in Mexico. The machine was reported to be worked good for crops like tomato, cucumber and green pepper. The machine was consisted of cleaning and waxing rollers. In the upper part of these rollers, a wax deposit was placed, which was kept liquid with the help of heat generated by light bulbs. Guides were provided to control the aperture of rollers as a function of size of the fruits. Once the fruits were cleaned and waxed, they go to the flat belt conveyor where they were grouped together according to their size, color, and external defects. The belt was 10 m long and 80 cm wide which lay on top of several smooth rollers. The machine was driven by a 4.48 kW electric motor.

Wardowaski *et al.* (1986) reported that, whatever the function of individual components is, a packing line is considered as of so many square meters as the conveying surface. The wider and shorter the line, the lesser the initial cost. This is so because (1) The length of the line largely determines the size of the building (Width makes very little difference to the building, size); (2) the sides of the line (framing, bearings, chains, drives, etc.) account for more to the initial cost than does the span; (3) electrical and plumbing costs relate more closely to the length than to the width of the line. The wider the line the slower the linear speed needed for a given capacity. Extra capacity can only be obtained by running fruit too fast, or else by building an additional line. When fruit treatment time is critical, the line can be pulled apart, and extra sections can be added at a less cost. Within reason, therefore, start with a line wide enough to accommodate the volume of fruit anticipated for the following 4-5 years.

Grierson and Wardowski (1973) reported that the fruit that is not going to be shipped as fresh fruit should be removed as soon as possible, especially when only one grade is packed. Immediately after dumping and trash elimination, the over and under sizes can be removed. There is no point in putting fungicides, waxes, and possibly color-add on fruit that are to be sent to the cannery. This costs the packinghouse money and gives the cannery problems. Grading immediately after the washer considerably reduces the amount of machinery necessary between the washer and the sizer. If 30-40 per cent of the fruit is being graded out to the cannery, the size and horsepower requirements of the remaining line can be decreased proportionally. The machine cost per box is decreased by grading immediately after washing.

Several operations are indicated as being time-dependent. The operations such as dumping, trash elimination, and pre-sizing do not involve retaining the fruit for a given length of time. Time-dependent processes, on the other hand, take a certain interval of time, which must be taken into consideration in calculations to determine the necessary area of conveying surface.

Dumping

Wardowaski *et al.* (1986) reported that if the machinery is well designed with unimpeded fruit flow, the fruit will suffer no appreciable damage. Dry dumping is also recommended for all soft citrus cultivars. If such fruits are allowed to soak in a water bath or fungicide solution their fragile and puffy rinds absorb water, making them even more prone to injury. Some packinghouses prefer to use wet dumping of fruits. Where wet dumping is practiced it is important to remember that, unless suitably treated, the water in such a bulk bath will accumulate decay-causing mold spores and act as an ideal source of infection for injured fruit. To avoid this, the usual practice is to add chlorine gas or powder (calcium hypochlorite) to give a concentration of 100-200 ppm active chlorine. The pH of the water must be maintained between 7.0 and 9.0.

Grierson (1971) in his study on packing house operations reported that few packing houses have provision for coping with the trash that is commonly included in deliveries of hand-picked fruit. A sloping belt trash eliminator can be used to remove the loose trash. A vacuum brusher can also be used to remove loose sand, which is extremely injurious to packing line equipment. Roller conveyors should always be self-cleaning, i.e., have angled supports that wipe the whole length of each roller with openings to shed trash. Several pre-graders with clippers are sometimes needed to remove any adhering long stems, and provision needs to be made for removing any rots and splits. Dawson and Eckert (1977) reported that where fruit is dry dumped, it is essential to wet it as soon as possible to avoid injury to the fruit as it passes over cleaning brushes; this is particularly important in the case of soft citrus cultivars with their sensitive rinds.

Washing/Cleaning

Use of transverse rotating brushes for washing is a common practice. The foam curtain should be located immediately above the space between the first and second brushes, but not immediately over a brush, because fruit pauses between the brushes. A minimum exposure time of 20s on the brushes is necessary for adequate washing; 30s is much better where fruit have considerable sooty mold. There is evidence that slower brush speeds are more efficient than faster speeds (Rejimbal and Bigler 1973). The recommended rpm for washer brushes are 120-145. Thorough rinsing is necessary after washing and is best done over the last of the washer brushes or over a roller conveyor. Each fruit should pass under at least two nozzles and the rinsing water should be supplied at a pressure of approximately 200 kPa.

Rejimbal and Bigler (1973) investigated the variables affecting the performance of brush washer equipments. Washers containing 26, 36 and 46 brushes were studied. The variation in soil adhered to fruits did not appear to affect brush washer operation. Adequate rinse spray application was found to be the most important single variable under either condition. Kaplan (1985) reported that water elimination sections typically contain sponge rubber do-nut rolls with squeeze rolls mounted on their underside to wring the water from the donuts. The wringer rolls should be under spring tension for efficient operation and for optimum results; the wringer rolls themselves should be wiped by a stationary blade. The sponge rolls can be supplemented by or replaced with soft brushes and angle iron flicker bars positioned beneath the brushes at a 45° slant on the downward movement of the brush. In this way, water will be flicked at a 90° angle into the drain pan and not onto the following brush. Since, water eliminator sponge rolls grip the fruit more firmly than brushes, their speed should be slowed to about 80 rpm, while the brushes rotate at 110 rpm. Similar method was also reported by Wardowaski *et al.* (1986) with exception that the roller speed should not exceed 75 rpm.

Hagenmaier and Baker (1993) reported the affect of cleaning methods on shrinkage rate of citrus fruits. The shrinkage rate of Marsh grapefruits and Valencia oranges was increased 50 to 150% by washing the fruits with rotary brushes, but was not changed by hand-washing the fruits with cellulose sponge. Thus, controlling washing process is important to minimize shrinkage rate of fresh citrus fruits.

Oh (2005) patented a method and equipment for cleaning fruits. The equipment includes a brush roll driving unit, a blower and a dust collecting system.

Degreening

Citrus cultivars mature with change in chemical composition-sugars-acid ratio and flavor, although the peel is still green, which is unacceptable to the consumer who usually associates green color with lack of ripeness. To improve peel color and to remove chlorophyll from the green peel, degreening treatment with ethylene gas is commercially applied in most citrus growing countries. But, degreening accelerates the senescence of the fruit by increasing respiration rate, transpiration, dryness of button ends and amount of decay. Therefore, determining factors for different citrus varieties at growing country is necessary (Cohen, 1999).

Ethylene concentrations: From 5 to 100 ppm ethylene causes no apparent difference in fruit color development. Higher ethylene concentration accelerates the initial disappearance of green color from the peel during the first hours of the process only. The higher the ethylene concentrations used, the higher the respiration rate was, and the faster the buttons dried and dropped. Direct correlation was found between degreening and stem-end rot developed caused by *Diplodia* and *Alternaria*.

Duration: The duration of ethylene process was more important for peel color than ethylene concentration. Shortening the degreening period by using higher concentration of ethylene cannot be achieved

Temperature: Temperature has a direct effect on loss of green color-breakdown of chlorophyll, and on the development of orange color-synthesis of carotenoides. Optimal recommended temperature is 25°C. Lower temperature will delay the disappearance of chlorophyll. Higher temperature than 28°C is desirable for chlorophyll destruction in green fruits as grapefruit and lemon, but will disturb carotenoides biosynthesis in orange color fruits such as oranges and mandarins.

Relative humidity: The relative humidity between 70 to 90% on color development throughout the degreening process had no effect. However, high RH prevent water loss and desiccation of the fruit peel. Peel breakdown, and peel disorders may be reduced with high RH during degreening. High RH may assist healing of damaged peel, thus reducing mould decay.

Air composition: The atmosphere of the degreening room is of course greatly affected by the rate of ventilation during the degreening. Ventilation is required to supply oxygen needed for ethylene to induce color changes rather than to remove CO_2 from the degreening room. Under the conditions pertaining in commercial practice rooms having high or low ventilation, after a similar period of degreening about same CO_2 was found. At these conditions there were no differences in color development, respiration and water loss, sugar and citric values in fruit juice, peel blemishes and decay response. Ventilation rare of 0.5 air changes per hour is recommended which greatly reduces the energy requirements for heating, saving ethylene gas, maintaining high relative humidity in the degreening rooms, and resulting improved degreened fruit quality.

Intermittent degreening method: It is a modification of the commercial degreening method based on intermittent exposure of the fruits to ethylene and heating: ethylene at 5 to 10 ppm, and heating to 25°C for a cycle of 12 hours, followed by a 12h interruption. The process lasted 96 h for dark green fruits at the early season, and 36 h for green-yellow fruits at late season, same as in continuous degreening method (gas and heating supply) without undesirable rises in ethylene concentration and temperature that sometimes occurred. Once color development started, it was unaffected by the 12-hours interruption. During the intervals, the decrease in ethylene concentration to 0-1 ppm and the temperature dropped by 1-2°C do not affect the color development. Relative humidity of 90% and ventilation rate of 0.5 air change (about 1% fresh air/min) are provided during the whole process. The economic benefit of the intermittent method is in saving ethylene gas and in energy. It provides good conditions for degreening and results in better fruit quality. Satsuma mandarin was degreened with the 3 common methods: the kerosenestove method in use in Spain, the 15-h ethylene treatment used in Japan and the intermi-ttent method' used in Israel. The intermittent method gave the best results for variable color, blemishes and rots.

Thiabendazole (TBZ) and 2,4-D treatment prior to degreening: Dryness of buttons and fall-out followed by stem end rots is common in ethylene degreened fruits. Dryness of buttons was related to a substantial increase in pectin methyl esterase (PME) and in cellulose activities in the abscission zone of treated fruit. While prior degreening treatment with TBZ controlled the *Diplodia* rot, the 2,4-0 preserved stem-end color vitality following reducing enzymatic activity. That maintained stem end green, and prevented rot fungi such as *Alternaria*. Only the 2,4-D amine formulation at 250 ppm and not the 2,4-D isopropyl ester who caused delay in color development, is recommended to be use with TBZ 500 ppm. The combined treatment is applied as a water drench prior to degreening.

APPLICATION OF COATINGS

Two types of waxes are used; in one form, the wax solids are dissolved in a petroleum-based organic solvent and fogged on rollers in a chamber run under slight vacuum to minimize fire hazard. The solvent waxes have been virtually eliminated due to environmental problems associated with evaporating the solvents into the atmosphere. Overhead fans, radiant heaters, or dry tunnels complete the preparation of the fruit for the solvent wax applicators. Heat lamps and fans avoid the hazard of heat damaging the fruit pulp. The rollers delivering the fruit to the waxer should have live rolls to aid in spreading the fruit. A solvent wax mist is formed by a blower nozzle assembly at the entry of the enclosed roll conveyor system. To aid in the distribution of wax and the adequacy of full fruit coverage, oil burner type nozzles should be flush with the air tube in which they are positioned. Nozzle pressure should be between 2.19 and 2.92 kg/cm^2. The blower fans should be set to rotate at about 1000 rpm, and air should be exhausted from the solvent wax enclosure at a rate of 61-67 m^3/minute. The air nozzles adequately atomize the solvent wax into the fine mist needed for good performance.

Kaplan and Petrie (1982) reported that the fruit entering the waxer should start turning on the second or third roller and should make at least three complete revolutions while in the wax mist for good coverage. The dwell time in the waxer should be at least 8s. The roller conveyor should be at least 90% full of fruit to avoid diverting the wax through openings between fruit. Incorporation of a variable-speed drive will permit adjusting the conveyor speed to match the fruit flow. The transfer of waxed fruit to the delivery board should be rapid enough to avoid fruit pinching at the transfer point. The conveyor carrying fruit away from the waxer should have non-rotating rollers that will not damage the wax film.

Long (1964) reported that the other type of wax application involves use of a water emulsion wax that can be applied as a dip, foam, or spray. The spray is the conventional method in most citrus-producing countries, the emulsion typically being applied over a bed of slowly rotating (100-120 rpm) brushes. For this application, horsehair is superior to plastic bristle, as its rough texture tends to hold the emulsion and wipe it on the fruit rather than throwing it off by centrifugal force. In some cases fine plastic bristles (Saran 0.009) can be used if horsehair brushes are not obtainable. The brushes should be of the corrugated or tumbler type so as to adequately expose all surfaces of the fruit to wax. The emulsion can be applied with fixed nozzles, but the traveling nozzle system ("wig-wag") is usually preferred. In this system, an oscillating nozzle emits a fine fog of emulsion.

Wardowaski *et al.* (1986) reported that in any wax application operation it is important to apply a uniform and controlled coating of wax to the fruit. This is particularly important where decay-control fungicides are added to the wax. The major cause of uneven wax coatings is variation in the flow of fruit into the wax applicator. This problem can be solved by the installation of an automatic control unit that monitors the flow rate of fruit as it approaches the waxing unit and adjusts the wax flow rate accordingly. Another, but less efficient, method is to maintain a constant wax flow rate but to automatically adjust the speed of the elevator/conveyor feeding the fruit into the wax applicator in accordance with the volume of fruit going through. Also, intermittent (pulsating) systems have recently been installed to avoid applying excessive wax. Unless precautions are taken, a brush bed used for application of wax, fungicide, or both, can contaminate fruit being run over it long after the changeover has been made. Very efficient cleansing of the brush bed can be obtained by the use of "flick bars" under the brushes. These are on a sliding rack under the horsehair brushes of the applicator.

The waxing unit should be designed for easy cleaning. Removing wax buildup in the waxing unit and along the exit conveyor is a mandatory chore for the end of each packing shift. Self-cleaning rolls should be used whenever possible. Wax residues on equipment can readily damage fruit and destroy the integrity of the wax coating. To clean an applicator, first rinse out the lines and nozzles thoroughly. There is no point in cleaning the brush bed and then flushing out the nozzles onto the brushes. Then start the brushes and pull flick bars back so each is in contact with its equivalent brush. Hose down for at least a minute with the flick bars in place and the brushes revolving. After hosing, leave flick bars in place for another 15-20s to dry off brushes and avoid diluting the wax (Wardowaski *et al.*, 1986). Finally, push flick bars into the neutral position and start up with the next lot of wax.

Anonymous (2001) reported that the performance of citrus wax is dependent on the method of application. The amount of wax applied and the uniformity of application are extremely important. Fruit should be damp dry prior to wax application to prevent dilution. Wax should be applied at full strength at the rate of 1.5 litres per tonne of fruit. Bureau of Indian Standards (1974) recommends wax application at the rate of 140 mg per kg of fruit. Waxes should never be diluted with water. Wax should be sprayed or dripped on to the fruit on a bed of brushes. The brushes should have at least 50% horse hair to help spread the wax over the fruit. A minimum 8 brushes should be used. Brush speed should not exceed 100 rpm (Anonymous, 2001).

Spinning head is a low pressure application system that can be used to put an even distribution of wax for application by brushes.

Singh *et al.* (2002) reported that in mechanized pack houses the sorting, washing, waxing drying, size grading, and packing of citrus fruits is done with machines. Sorting is done manually by removing the culled or undesirable fruits by the trained persons while fruit is moving on the roller conveyor. In automatic system fruits are sorted by electronically controlled photo-electric cells on the basis of color and rind blemishes. Washing is accomplished by spraying water through nozzles while fruit is moving ahead on nylon roller brushes. After washing, the fruits are coated with wax emulsion applied through atomizer nozzles with air pressure. Wax is coated on entire fruit surface by horse hair brushes. The fruit coated with the wax emulsion then passes through a dryer tunnel. The air temperature is raised up to 55 to 60°C and hot air is forced through the coated fruits moving on roller conveyor. Fruit coating dries within one and half minute

Wagner and Sauls (2004) reported that water emulsion waxes do not require complete surface drying of fruit, so the drier is unnecessary. The wax is applied as foam, dip, spray or brush wipe. Spraying is most common, using a pair of traveling nozzles (wig-wag) over a bed of slowly rotating (not more than 100 rpm) horse hair brushes. Brushes can be cleaned with flick bars on a sliding rack between the brush bed. Waxed fruits are dried on a "wax-setter", which is a two-sectioned slat conveyor, 6.1 to 12.2m long. The outlet of the first conveyor is a few inches above the inlet end of the second one so that the fruit will flip over as it moves onto the second conveyor.

Surface drying

Fruit surface drying is normally done with hot air, which used to be delivered over fast-moving polisher brushes. The combination of hot air and brushing can be a source of considerable damage to tender fruit, and is discouraged. Where brushes are used, they should turn at no more than 100 rpm, and then preferably only with cold air. Miller (1978) reported that the maximum moisture at the dryer inlet was about 1.68 g per fruit for Marsh grapefruit on a per fruit basis.

Wardowaski *et al.* (1986) reported that drying is a time-dependent operation. The time necessary depends on whether wet fruit is being dried before waxing, in which case 1.75-2 minute is usually adequate. When wax is being dried on the fruit, 2.5-3 minute (or even longer if ambient temperature air is used) is usually necessary to dry a water wax and about 1 min for a solvent wax. Dryers for waxed fruit should always have a short initial section and a longer final section. This design largely confines wax accumulation on the equipment to the first short section. Waxed fruit should not roll while drying. This is usually achieved by using either a slat conveyor or, preferably, "stilled" rollers, i.e., rollers supported by the pin chain at the sides so that they do not turn.

The temperature of the hot air that can be used without damaging fruit depends on several factors. The first of these is the fragility of the fruit. Tender speciality fruit, such as degreened tangerines and other soft citrus types, should be heated as little as possible and not brushed at all. Midseason grapefruit seldom suffer damage during drying. Wet fruit will stand very much higher air temperatures than dry fruit. Valencia oranges subjected to 66°C air showed no increase in pulp temperatures as long as the fruit surface was still wet (Grierson, 1957). However, temperatures above 54°C are not recommended, particularly at the downstream end of the dryer where the fruit are dry or nearly so. In view of the sometimes detrimental effect of hot air in the drying process and the relatively high cost of producing it, consideration should be given to an alternative technique. As drying is very much a time-dependent factor, the longer the fruit remain in the dryer, the better it is. If a conventional dryer elevator is slowed down to the extent that the fruit pile up to form two or three layers (so-called multiple layering) the through-put capacity of the dryer will not be reduced by the extra time (increased from 6 minute to 9-12 minute). With the extra drying time achieved, it has been found that ambient temperature air blown down through the fruit by powerful overhead fans is sufficient to obtain good drying. With this method it is also not necessary to enclose the dryer unit.

Wardowaski *et al.* (1986) reported that very rapid drying is possible with infrared rays. The infrared effectively dries a band around the fruit, but not the ends of the fruit. This makes it suitable for pre-drying; after infrared exposure, the fruit no longer wet the rollers as they move into the conventional dryer. Metal absorbs infrared, so metal rollers cannot be used, and all metal surfaces exposed to the rays must be shielded wherever they might touch the fruit and cause a severe peel burn.

The surface drying of fruits after the application of wax coating can be achieved using driers consists of wide open conveyors with vertical fans mounted above the fruit flow, with larger units consisting of heater fans spaced alternately with cold fans. The drying process is a time dependent operation with approximately 2.5 to 3 minute required to dry the wax (Anonymous, 2004).

Grading

In general, blemished fruit are removed and the desired grades are left on the line. It is common practice to grade fruits on a simple roller conveyor, although various improved graders have been devised (Bowman 1975, Hunter *et al.,* 1958). In pre-grading which is done before washing, fruits for canneries are sorted out. In the second stage, after waxing but before sizing, fruit for the local market is sorted out, then sized and packed on a separate line. The final grading is normally done on belts moving on an inspection-type table where the fruit from each grader can be inspected. The main flow is mechanically divided into two sizes, large and small. The flow of fruit in each size is then divided and fed to groups of three or four graders. These graders separate the flow into the various grades by removing the undesirable grade and leaving the desired grade on the belt

Electronic sorting machines are technically capable of sorting for both color and blemishes (Gaffney, 1973). It has been suggested that the best prospects for economic use of such machines be in a highly mechanized, combined fresh fruit and cannery operation (Grierson and Wardowski, 1977).

The feasibility of electronic sorting for surface defects has been thoroughly investigated in Israel (Felsenstein and Manor, 1973) for fruits to be marketed in Europe and Japan. It was found that technically the detection of most blemishes is possible, but the processing of data, in order to decide whether the fruit should be rejected or accepted, is still to be investigated. Also, at present there is no economic justification for the development and incorporation of an electronic blemish-sorter, since not all blemishes can be sorted out mechanically and a significant amount of hand sorting will also have to be done. The choice of sizer depends on various factors such as shape of fruit, volume and rate of packing required, and size of packinghouse.

Allhouse and Stephenson (1969) reported that the most common types of conveyors for handling individual fruits or vegetables are flighted or cup belts. However, both types present problems when each individual fruit or vegetable must be handled separately from all others. Unless belts are narrow, the product can roll from side to side thereby causing unnecessary damage. Unless flights are high, the product can roll back over the flight; in that case the product is no longer singularized. Also, the narrow width of the conveyor entrance can cause jamming when more than a single fruit or vegetable attempts to enter the same channel at the same time. Deflecting vanes and similar devices at the entrance may not remedy the problem.

Stevens and Gale (1970) studied the feasibility of semi-automatic quality inspection of fruits and vegetables. The equipment was an inspection table consisted of roller conveyors. The rollers were of 5.6 cm diameter with 10 cm spacing between centers, the width of the rollers was 45 cm divided into four equal lanes. Operators were seated either facing the table or sideways facing the apples coming from their left, and they were provided with a keyboard on which to register their assessment of the grades.

Goodman and Hamann (1971) reported a machine to size the sweet potatoes on the field itself. The sizing of the roots accomplished by having them ride in a 'V' which had an open bottom that gradually increased in width from the front of the unit to the rear. The sides of the 'V' were made up of round, clear vinyl plastic belting driven by step-cone sheaves which were keyed to a one-inch shaft mounted in pillow block bearings and driven by roller chain from hydraulic motor. The belt near the top of the 'V' has a higher linear speed than those near the bottom. The effect of this was to cause any object, such as root, that fell on the sizing belts to rotate until its long axis was parallel to the belts. The machine was tested for sizing accuracy, and the damage caused to the roots.

Peleg and Ramraz (1975) reported main reasons for sizing fruits as ability to market pattern packed, uniformly graded fruit to enhance "sales appeal"; the ability to fill shipping containers by count, which is faster and cheaper than filling by weight; and ability to mechanize pattern packing of fruit in shipping containers. The sizing parameter of a sizing machine is a function of its construction that is the principle of sizing it employs. As long as the machine is in good condition its sizing parameter function remains fairly constant, provided fruit properties are unchanged. Sizing accuracy of a sizing machine may be defined as the percentage of times it errs, when sizing a typical fruit sample i.e. what percentage of fruits ends up in the wrong counts. Numerically sizing accuracy is best expressed by the "Mean sizing error" (M) defined as the mean percentage of a typical fruit sample outside the separation lines of their respective counts, on the sizing map.

Hood *et al.* (1981) developed a fruit handling system which consisted of tractor and a bin trailer. A 37 kW tractor equipped with a front loader frame was used to power the system. A hydraulically powered side mounted lift conveyor on the tractor receives fruit that are discharged from the mechanical harvester. The conveyor was 427 cm long with an inside belt width of 42 cm. the trailer was made to accommodate four pallet bins (122 cm X 122 cm X 61 cm) supported on conveyor rollers. Bins enter and exit the trailer at the back by use of a fork lift and/or the hydraulic cylinder fitting feature that allows filled bins to exit onto a concrete slab. Trailers of similar design and of four and larger bin capacities are currently in use. The study demonstrated that a tractor trailer fruit handling system has potential for collecting fruit from mechanical harvester.

Shyam *et al.* (1990) reported the design and development of potato grader. The emphasis was that the machine should be simple in construction and capable of sorting potatoes into 4 or 5 different sizes or grades with high sizing efficiency and 2 to 2.5 t/h capacity. The machine consists of a steel frame, power transmission system, a set of two ossicilatory sieves, a stationary sieve, feeding chute, sorting platform and bag filling device. A single phase, 0.746 kW electric motor of 1440 rpm with speed reduction ratio of 6.25:1 was used to operate the sieves. Screen efficiency of the oscillating sieves varied between 80 to 90%. It requires 10 to 14 attendants for the operation.

Anonymous (2004) reported that the citrus fruits are graded by size. This can be done by hand or by machine. If the grower is grading citrus manually then its best not to judge the size only by eyes, but to use some kind of measuring device. A simple way to check the fruit size is to cut a series of round holes in a thin wooden board or a piece of thick card board, according to standard market size for that variety.

Chapter 4

DEVELOPMENT AND EVALUATION OF PROTOTYPE

There is no dearth of advanced technologies available today for postharvest handling of fruits and vegetables. But, still there are countries like India where postharvest losses of these commodities are much higher than the developed countries. The buying of advanced technologies from the developed countries at exorbitant prices is not a feasible option with the countries like India where the average farmers are already debt ridden. The best way for such countries is to evolve low cost machineries with acceptable efficiencies. Such machines will be welcomed by the farmers.

CONCEPT BUILDING

Discussions were made with some progressive Kinnow growers; Joint Director Agriculture (Agriculture Engineering, Chandigarh office), and related Assistant Agriculture Engineers, Department of Agriculture, Punjab, as and when required during the concept building process. Visits were made to the existing Kinnow waxing units run by Punjab Agri Export Corporation (PAGREXCO) at Tahli Wala Jatan (District Firozpur) and Chhauni Kalan (District Hoshiarpur) to study their various mechanisms and to identify the areas of cost cutting without affecting the performance of the mechanism. Discussions were also made with the persons attending these units regarding the performance of existing units and, need and scope of low cost on-farm cleaning, waxing, and grading units. Discussions were also made with the experts working in the field of postharvest handling of fruits and vegetables in the Department of Processing and Food Engineering, to finally work out the theoretical design and fabrication details of the mechanism whose performance quality could be accepted for the market conditions and whose acquisition cost could be adequate to the economical possibilities of an average Kinnow grower.

DESIGN AND DEVELOPMENT OF PROTOTYPE

The available information suggested that the waxing and grading machines generally consists of cleaning and dewatering unit, waxing unit (coupled with surface drying unit), and the grading unit. Following this pattern the prototype was divided into four sub-units such as (1) cleaning unit, (2) waxing unit, (3) surface drying unit, and (4) grading unit. Each of these sub-units was designed and fabricated to accomplish the concept of on-farm cleaning, waxing, and grading of Kinnow fruit. The following functional design requirements of the prototype were considered:

- It should be simple and easy to operate
- Cost of the machine must be affordable to an average Kinnow grower
- It should meet the basic requirements of an efficient cleaning, waxing and grading machine

To develop the sub-units of the prototype the characteristics of Kinnow fruit such as size, shape and angle of repose were considered essential. The average size of Kinnow fruit was measured as 65 mm. The angle of repose of the Kinnow fruits was measured and depending upon the position of large and small axis of the fruit it was found to vary between 4° and 19° for an average sized fruit.

As mentioned earlier in chapter 3 that several operations are indicated as being time-dependent. The operations such as dumping, trash elimination, and pre-sizing do not involve retaining the fruit for a given length of time. Time-dependent processes, on the other hand, take a certain interval of time, which must be taken into consideration in calculations to determine the necessary area of conveying surface. Surface drying is a time dependent process in the whole waxing and grading mechanism (Wardowaski *et al.,* 1986, Grierson 1957, Anonymous 2004). Thus, it determines the working parameters such as linear speed/rpm, width, and hence, the capacity of the machine. Also, it was reported that, whatever the function of individual components is, a packing line is considered as of so many square meters as the conveying surface. Therefore, it was decided to first calculate the dimensions (length and width) of conveying bed in the surface drying unit. The dimensions of time independent operations can be calculated to match the overall capacity required. The dimensions of cleaning and waxing sub-units also depend upon the minimum number of rollers required in these units.

Dimensions of surface drying unit

The drying unit's generally consists of a conveying assembly, air blowing system, and the system to heat up the ambient air. For waxing and grading units the tunnel type dryers are commonly used. The machines capacity depends upon the forward speed and the dimensions such as length and width of the conveying assembly. For a given capacity the forward speed and hence, the length and width of the conveyor can be calculated using Eq 4.1.

$Q = 3.6 (q \times v)$ ....Eq (4.1) (Spivakovsky 1985)

Where

Q = Capacity (t/h)

q = Weight of the fruit per meter length of the conveyor

v = Linear speed of the conveyor (m/s)

The residence time of fruit inside the surface drying unit was assumed to be 2.5 minutes. For 0.5 t/h capacity (determined on the basis of average orchard size) the linear speed, hence the length of the surface drying unit was calculated using Eq. (4.1).

The number of average sized Kinnow fruits per meter length was measured to be approximately 15. Due to high fabrication cost of the roller brushes (Fig. 4.1) to be used in cleaning and waxing unit, the width of conveyer in surface drying unit was restricted to 475 mm. Assuming 90 per cent loading efficiency, the effective width of the conveyor comes out to be 425 mm. The weight of Kinnow fruit per meter length of the conveyor comes out to be approximately 4.65 kg. For 0.5 t/h capacity the linear speed was calculated using Eq. (4.1) as 0.03 m/s. Taking residence time of fruit inside the surface drying unit as 2.5 minute, the length of the surface drying unit comes out to be 4500 mm.

ESTIMATION OF SIZE AND MATERIAL OF PROTOTYPE COMPONENTS

Having decided the dimensions of the time dependent operation, the dimensions of other sub-units were determined to match the overall capacity of the prototype. The unit-wise size determination and the estimation of material requirements are discussed in this section.

Cleaning and waxing unit

The cleaning and waxing operations normally accomplished by revolving brushes mounted on a frame. The dimensions of these units are determined by the size and number of the brushes to be used. The number of brushes also determines the quantity of other components required. On the basis of information reported (Anonymous 2001, Kaplan 1985) the minimum number of roller brushes required in cleaning and waxing units are 10 and 8 respectively. The cleaning and waxing is accomplished through the brushing action of rollers while the fruits roll in between the troughs made by two consecutive rollers. Therefore, taking average size of fruits as 65 mm the diameter of the rollers was fixed as 100 mm. Taking care about the matching of various sub-units and the cost of roller fabrication into account the length of rollers was fixed as 330 mm. Considering various allowances for assembling of different parts such as bearing and sprockets etc., together and smooth functioning of the sub-units; the length, width and height of main frame of cleaning and waxing units were worked out. The dimensions of the main frame of cleaning unit were determined as 2300 mm x 435 mm x 1530 mm; and for waxing unit the dimensions were 1400.4 mm x 435 mm x 1215 mm.

Fig 4.1 Roller brushes used in cleaning and waxing units

The number of shafts required for fixing the rollers on main frames was equal to the number of rollers. The number of bearings, bearing brackets, and the number of sprockets required were double the number of shafts to be fixed. The number of nuts and bolts required was four times the number of shafts. The size of the nuts and bolts was fixed by the size of the holes provided in the bearing brackets.

To avoid side spillage of fruits from the rollers, side supports were provided along the length on the two sides of the cleaning and waxing units. Due to the motion of fruits along the length of the conveyer, the side thrust of the fruits was taken negligible therefore; the side supports were made out of 24 gauge GI sheet.

For power transmission in cleaning unit an electric motor, two pulleys, and a belt were required. For waxing unit, the power drive was given from surface drying unit, therefore, angle iron, a shaft, two bearing and bearing brackets, two pulleys, a belt, and a sprocket were required.

Surface drying unit

The material requirements of different components of surface drying unit such as conveying system, electrical system, and drying tunnel depends on the size of the conveying bed.

Size of main frame: The dimensions-length and width of the main frame of surface drying unit is the same as that of the conveying assembly as determined in section 3.3.1. The height of main frame of the unit was decided on the basis of the height of the rolling bed in waxing unit, in such a way that, while transferring the Kinnow fruits from the waxing unit to the conveying bed the fruits should roll down without any injury. The height of the main frame of the unit was determined to be 1075 mm. Therefore, the dimensions of the main frame of surface drying unit came out to be 4500 mm x 42.5 mm x 1075 mm.

Conveying system: To convey the fruits in surface drying unit, a conveying bed made from plastic PVC pipes fixed on the two parallel chains was used. A feeder box type chain that is used in combine harvesters to convey the material to threshing drum from auger of the machine was selected. Local manufacturing, easy availability, and its low cost were the factors behind the selection of this chain. Total 19400 mm long chain was used in the conveyer.

Electrical system: The electrical system comprises of heating load, temperature controllers, circuit breakers, and blowers. The material estimation for the electric fittings of the unit was made on the basis of the design of drying tunnel. Design of drying tunnel and material estimation was based on the following information available in the literature:

- The maximum temperature inside the drying tunnel should not exceed 45°C
- The residence time of fruit inside the drying tunnel should be 2.5 minutes
- Based on the meteorological data of Ludhiana city for the last 10 years the average ambient temperature during the months of December, January and February were considered as 14.3°C, 10.7°C, 15.3°C, respectively; and the average relative humidity during these months were considered as 69, 80, and 68 per cent respectively.

The heating load for surface drying of fruit was calculated to be 7 kW. The details of calculations are given in Appendix B. For easy monitoring of temperature, it was decided to provide fourteen heating elements each of 500 W rating. As suggested in literature (Anonymous 2004), it was decided to blow the air in cross flow pattern. To this effect 8 blower, each of 300 mm diameter was used. To monitor temperature inside the drying tunnel 3 thermocouples, and 3 temperature indicators were used. The temperature inside the drying tunnel was controlled using 3 circuit breakers. Fourteen switches for heating elements, and eight speed regulators-cum-on/off switches for fresh air fans were used.

Power requirements: Keeping in view the fact, that, the limiting friction is always greater than the rolling friction, therefore, the torque required to rotate the conveyer shaft was measured with the help of torque indicator available in the department of Farm Power and Machinery, PAU, Ludhiana. To compensate for any error in the measurement, a factor of safety of 100% was considered. The combined power requirements of conveyer in surface drying unit and waxing unit comes out to be 0.746 kW. The power from a d.c. motor was transmitted to the conveyor through belt and pulley arrangement.

Drying tunnel: The casing of drying surface unit was made from GI sheet. For fixing GI sheet, and the electric fittings a frame of angle iron was made. The frame was welded to the main frame of the unit.

Grading unit

A number of variations in grading unit, from simple to highly sophisticated, design are available. Keeping in view the low cost and simplicity of operation it was decided to make some mechanism, such that, the grading of fruits does not require any moving parts. Therefore, the grading unit was designed on the principle that, the fruit will fall through the cavity of its size while rolling down on two inclined bars. The fruit fall was facilitated through force of gravity by giving an inclination angle of 9° with the horizontal. The length of the grading unit was decided on the basis of a preliminary experiment conducted to determine the number of revolutions a fruit needs prior to fall through the cavity while rolling down the inclined plane. On the basis of this experiment and taking allowance to eliminate any error in the design, the length of the grading unit for 6 different grades of the fruits was calculated to be 1420 mm; and for two row grading unit the width came out to be 400 mm.

The frame of the grading unit was made from angle iron whereas the diverging bars were made from GI sheet.

On the basis of information given above about the various sub-units, the requirements of materials for 0.5 t/h capacity on-farm waxing and grading machine is given in Table 4.1.

FABRICATION OF PROTOTYPE

The assembling of various parts into sub-units and sub-units into complete prototype was carried out in the workshop of the Department of Processing and Food Engineering. The machining of various components was done in the Research Hall of College of Agricultural Engineering, PAU, Ludhiana. The sequence followed for development of prototype was - cleaning unit, waxing unit, surface drying unit, and the grading unit. The material listed in Table 4.1 was used to fabricate the prototype.

Cleaning unit

The fabrication of cleaning unit was accomplished by mounting the roller assemblies on the main frame of the unit. The roller assemblies were made by drilling a hole along the axial line, at the center of the wooden rollers. The mild steel (MS) shafts of 25 mm diameter were sized to fit inside the hole drilled in the rollers. To accommodate bearings and sprockets for power transmission, the length of the shafts were kept 200 mm more than the length of the rollers. The cleaning unit which consists of 10 roller brushes is shown in Fig. 4.2. On both sides of the shafts, roller bearings were fixed. The matching bearing brackets were used to provide proper seat for the bearings on the frame of the cleaning unit. These bearing brackets were fixed on the main frame with nuts and bolts. The rollers were fixed at a center to center distance of 100 mm. Below the battery of roller brushes a rectangular plate of GI sheet of dimensions 1270 mm x 450 mm x 50 mm was placed to collect the dirt etc. The main frame of cleaning unit was made of angle iron.

Fig 4.2 Arrangement of rollers and power transmission scheme used in cleaning unit

The cleaning unit was powered with a d.c. motor fixed at the discharge end of the unit. Power from the motor was transmitted to the one of the roller brush with the help of belt and pulley arrangement. Power from this roller brush to the other brushes was transmitted with the chain and sprocket arrangement.

Table 4.1 Material requirements for fabrication of on-farm waxing and grading unit

Name of sub-unit	Material required	Specifications	Quantity
Cleaning unit			
1	Angle iron	50 mm x 50mm x 6 mm	80 kg (appx.)
2	Roller brushes	Diameter 100 mm Length 330 mm	10

3	MS round	Diameter 25 mm Length 530 mm	21 kg (appx.)
4	Bearings	Internal diameter 15 mm	20
5	Bearing brackets	Matching with bearings	20
6	Sprockets	Pitch 9.525 mm	19
7	Roller chain	International number 06B-1	2 standard lengths
8	Pulleys	Diameter 100 mm	2
9	Belt	A type, 86 number	1
10	Electric motor	0.746 kW d.c., excitation 230 V 220 V, 50 Hz	1
11	GI sheet	Gauge 24	3 kg (appx.)
Waxing unit			
1	Angle iron	50 mm x 50mm x 6 mm	60 kg (appx.)
2	Roller brushes	Diameter 100 mm Length 330 mm	8
3	MS round	Diameter 25 mm Length 530 mm	20 kg (appx.)
4	Bearings	Internal diameter 15 mm	16
5	Bearing brackets	Matching with bearings	16
6	Sprockets	Pitch 9.375 mm	15
7	Roller chain	International number 06B-1	1 standard lengths
8	Pulleys	Diameter 100 mm	2
9	Belt	B type, 84 number	1
11	GI sheet	Gauge 24	8 kg (appx.)
12	Sprayer	Ultra low volume ASPEE make	1

Surface drying unit

1	Angle iron	50 mm x 50mm x 6 mm	120 kg (appx.)
2	Angle iron	25 mm x 25 mm x 4 mm	30 kg (appx.)
3	Square pipe	25 mm x 25 mm	30 kg (appx.)
4	Heating elements	Type U Length 300 mm Rating 500 W	14
5	Blower	Diameter 300 mm	8
6	Temperature sensor	PT 100	3
7	Temperature indicator	0-400°C 96 mm x 96 mm	3
8	Circuit breaker		3
9	Electric switches	1000 W	14
10	On/off switch-cum-regulators	7 speed	8
11	Electric wire		20 m
13	GI sheet	Gauge 24	40 kg (appx.)
14	Acrylic sheet	4 mm	4.5 m x 40 cm
15	PVC pipe	Diameter 25 mm	18 standard lengths
16	MS round	Diameter 50 mm Length 550 mm	15 kg (appx.)
17	Bearings	Internal diameter 30 mm	4
18	Sprockets	Pitch 40.625 mm	4
19	Feeder chain	Pitch 40.625 mm	19000 mm
20	Electric motor	0.746 kW d.c., excitation 230 V 220 V, 50 Hz	1

21	Pulleys	Diameter 75 mm Diameter 275 mm	2
22	Belt	Type B, number 84	1
Grading unit			
1	GI sheet	Gauge 24	10 kg (appx.)
2	Angle iron	25 mm x 25 mm x 4mm	10 kg (appx.)
Nut – bolts and washers (Assorted sizes)			40 kg

An 'L' type side support 90 mm high and 45 mm wide at the foot made from GI sheet was provided on both sides of the rollers. The side support was fastened above the bearing brackets.

At the feeding end, to facilitate the feeding at constant rate a small chute was provided. The chute was placed just above the first roller, covering half of the roller. At the discharge end, a chute was placed just below the roller covering three-fourth of the roller. This chute also served as a feeding chute to the waxing unit.

Waxing unit

The waxing unit (Fig. 4.3) was fabricated in the manner similar to that of cleaning unit, with the exception that the bristles of the roller brushes were that of horse hair. In this unit 8 rollers (similar in dimension of that of cleaning unit) were used. The details of material, used such as angle iron, bearings, chain and sprockets, and power drive to each roller are listed in Table 4.1. Here also, side supports of GI sheet were provided in the fashion similar to that in cleaning unit. To avoid the drifting of wax spray during the operation the height of side support was kept 265 mm.

An ULV sprayer (Fig. 4.4) was mounted on the frame to spray the shellac based film emulsion on the fruits moving on the rolling brushes. The sprayer was placed on a frame made of flat steel bars. The frame was supported with the welded joints on the frame of surface drying unit 400 mm above the conveyer bed. The sprayer was positioned to spray the emulsion in the direction opposite to the motion of the fruit.

Fig 4.3 Arrangement of rollers and power transmission scheme used in waxing unit

Fig 4.4 Placement of ULV sprayer in waxing unit

The power to the brushes of waxing unit was provided from surface drying unit with the help of belt and pulley arrangement. A shaft of 25 mm diameter was mounted on a frame made from angle iron. A frame along with a shaft fitted on it with two bearing with bearing brackets, one sprocket, and a pulley, was welded to the frame of waxing unit 600 mm below the roller at the discharge end (Fig 4.5). The pulley was engaged to the driver pulley fixed on the driven shaft of surface drying unit with a belt. The sprocket was engaged to the driven sprocket fitted on one end of the roller brush at the discharge end of the waxing unit.

Fig 4.5 Power drive to waxing unit

Surface drying unit

The main frame of surface drying unit was made from angle iron. The conveying bed of the drying unit was made by fixing PVC pipes on the two parallel chains of pitch 40.625 mm. The pipes were fixed with the help of nuts and bolts. Two shafts of 35 mm diameter each were sized to fix two bearings of 30 mm internal diameter and two sprockets of 28 mm internal diameter on each of these shafts. These shafts were used as the driving and driven shafts on which the two parallel chains fixed with PVC plastic pipes were mounted. On the driven shaft (waxing unit side) a pulley was fixed to transmit the power to waxing unit. On the driver shaft (discharge end of unit) a belt and pulley arrangement was made to transmit the power from a d.c motor. The motor was fixed near the footings of the main frame.

The heating elements were placed 430 mm above the surface of the conveyer. The whole length of drying unit was divided to five compartments. The first, third and fifth compartments were providing hot air stream and the second and the fourth compartments were for cold air stream. The first and third compartment were provided with the 3 kW heating load, whereas, the fifth compartment was provided with the 1 kW heating load. The second and fourth compartments were not provided with any heating load. The blowers were fixed at 810 mm distance above the conveyer bed.

The surface drying unit was covered from the three sides with GI sheet. The feeding end and the discharge end of the unit were not covered. An inspection window was provided on one side of the surface drying unit. In the top covering holes were made to install the blowers. The covering sheets were fastened to the angle iron frame welded on the main frame of the unit. To insulate the top of the unit, a 4000 mm long, 230 mm wide and 230 mm high frame made of 25 mm square pipe was placed and was covered with 6 mm thick woolen sheet.

To monitor the temperature inside the surface drying unit three thermocouples were used to sense the temperature in the three compartments. These thermocouples were placed 270 mm above the conveyer bed. The thermocouples were connected with the three temperature indicators through the circuit breaker. Each temperature indicator was set for a maximum temperature required in that particular compartment. The arrangement of blowers, heating elements and the thermocouple is shown in Fig 4.6. The complete surface drying unit is shown in Fig 4.7.

Fig 4.6 Arrangement of blowers, heaters and thermocouples in surface drying unit

Fig 4.7 Surface drying unit

Grading unit

At the discharge end of the surface drying unit a rectangular chute of size 475 mm x 215 mm x 150 mm was attached to convey the fruits to the grading unit. The frame of the grading unit was made from angle iron. The three inclined bars forming two rows were made from GI sheet. The central bar was fixed and, the distance between the central bar and the other two bars were made adjustable by providing an oblong groove of 25 mm length. Fruits can be made into six grades ranging from 35mm to 85mm and above. To collect each grade an inclined plate of size 1210 mm x 555 mm x 50 mm, divided into six partitions, each of 30 mm height was provided below the inclined bars from where the fruits were routed to the collecting crate/boxes. The grading unit is shown in Fig. 4.8.

Fig 4.8 Grading unit

Keeping in view the smooth operation and space requirements the sub-units were positioned together in two parallel lines. In first line the waxing and surface drying units were aligned in a straight line. In second line, parallel to the waxing unit was the cleaning unit. The cleaning unit was positioned, such that, the hopper at its discharge end was made to serve as the input hopper for the waxing unit. The grading unit was positioned in-line with the cleaning unit and parallel to the surface drying unit. The discharge end hopper of surface drying unit was made to serve as input hopper for the grading unit. The positioning of sub-units in such manner helped reducing the length of the machine, thereby reducing the covered area required for the installation of the unit.

PROCEDURE FOR PERFORMANCE EVALUATION

The developed prototype was evaluated for its performance. Initially the prototype was run under no load conditions for 2 h. The desired adjustments were made to achieve the designed parameters such as linear speed and temperature etc. Later, experiments were conducted to evaluate the performance of prototype for cleaning, waxing and grading of Kinnow fruits. A trained person was employed to feed the raw material at a constant rate of 9 kg/minute (approximately one fruit per second). This is a minimum feeding rate that is required to maintain the continuous flow of fruits in the line. The machine was operated at designed speed 0.03 m/s. The performance of the developed machine was studied for wax application efficiency in terms of wax application rate per tonne of Kinnow fruits, and grading efficiency for various grades of the fruits. The surface drying unit was also evaluated for its working parameters and any changes required in the unit to accomplish the surface drying of fruits to the desired level were noted.

Cleaning unit

A minimum exposure time of 20s on the brushes is necessary for adequate washing; 30s is much better where fruit have considerable sooty mold. There is evidence that slower brush speeds (within reason) are more efficient than faster speeds (Rejimbal and Bigler 1972). Therefore, the performance evaluation of the cleaning unit was gauged from the time taken by the fruits to roll over the battery of brushes in cleaning unit. The weight of dirt collected in the collecting tray was taken. Total 10.5 kg fruit was cleaned to check the performance of the cleaning unit. Visual inspection was also made to check the fruits for their cleanliness.

Waxing unit

Wax application rate is very important for the successful operation of any waxing unit. Therefore, the wax application rate was determined. A known quantity of Citrashine was filled in the buffer storage tank of the sprayer. The sprayer was adjusted for wax application rate of 1.5 kg per tonne of fruits as recommended by Bureau of Indian standards (1974). The quantity used during the waxing operation was taken as the total quantity applied on the fruits. However, some formulation was deposited on the brushes which can be taken as the net wastage. This wastage was calculated from the difference of total formulation sprayed and the weight gained by the fruits during the waxing operation.

The weight gained by the fruits during waxing operation gave the quantity of wax applied on the fruits and hence the wax application rate.

Surface drying unit

To achieve the desired temperature inside the drying tunnel, the unit was run 15 minutes before the start of waxing operation. The blowers were switched on after 10 minutes of switching on the heating elements. The temperature indicators were set for the maximum temperature required in the different compartments of the drying tunnel. The first and third compartments were fixed for maximum temperature of 45°C, and the fifth compartment was fixed for 35°C. The temperature range in which the surface drying unit operated was recorded. Surface drying of the fruits at the discharge end of the drying unit was observed in terms of wet, sticky or dry. Mechanical damage caused to the fruits inside the surface drying unit was observed. For the smooth working of unit any other changes required in the design of the surface drying unit were also observed.

Grading unit

The grading efficiency was measured by grading the known quantity of fruits and repeating the exercise a number of times. Two fruits of three grades 46-55 mm, 56-65 mm, and 66-75 mm each were fed manually through the feeding chute 50 times. The number of times the fruits were graded into right or wrong grade was recorded. The grading efficiency was determined as the probability error in selecting the right grade.

PERFORMANCE EVALUATION

The developed prototype was evaluated for its performance in cleaning, waxing, surface drying and grading of the Kinnow fruits as discussed above in section 4.5.

Cleaning unit

The fruits took about 45s to roll through the cleaning unit. A minimum exposure time of 20s on the brushes is necessary for adequate cleaning in water washing system; 30s is considered much better where fruit have considerable higher level sooty mold (Rejimbal and Bigler 1972). At the end of cleaning operation, the visual inspection showed that the fruits were adequately clean. During experiment about 100 g dirt was collected in the tray placed below the cleaning brushes from 10.5 kg of fruits. As the fruits were harvested for waxing purposes, therefore, no other material such as leaves or stalks was there in the collecting tray.

Waxing unit

The waxing rate was measured as per the procedure given in section 4.5.2 The initial and final weight of the fruit to be waxed was 52 kg and 52.13 kg respectively. The waxing rate was determined as 2.5 g per kg of fruit. The recommended rate of wax application is 1-1.5 g per kg of fruit (Anonymous 2004, Bureau of Indian Standards 1974). The high rate of wax application may be attributed to the fact that the wax sprayer was placed to spray the wax on all the brushes of waxing unit, whereas in the waxing units in Punjab the wax is sprayed on the first two brushes only and subsequently the brushing action of other rollers help in uniform application of the coating (personal visit). Due to higher wax application rate than the recommended rate, the performance of the waxing unit in terms of wax application rate could not be considered as optimum.

No wax was collected in the tray placed below the waxing brushes. The quantity of wax deposited on the brushes during the operation was determined to be 166.4g. Therefore, this quantity can be considered as the minimum quantity of wax going waste during the operation. This surplus wax can easily be removed by rinsing the brushes with warm water.

Surface drying unit

The surface drying tunnel was divided into three heating compartments. The first heating compartment attained the temperature of 35°C within 15 minutes of switching on the heating elements. The side walls get heated up and worked as heat source afterwards and helped maintaining the temperature inside the drying tunnel to the required level. The blowers when switched on after 15 minutes of switching on of heaters helped to increase the temperature to 45°C by spreading the accumulated heat inside the compartment. The temperature drop of about 5°C was noticed inside the tunnel after 2 minutes of switching on of the blowers. This may be attributed to the fact that the blowers circulated fresh air continuously and the accumulated heat was not sufficient to heat up the air to maintain the temperature of 45°C inside the drying tunnel. The first heating compartment operates between 40-45°C.

The third heating compartment attained the temperature of 35°C within first 15 minutes of switching on of the heating elements. After switching on the blowers the compartment attained the temperature of 45°C. The temperature drop of 5°C was also noticed in this compartment of the drying tunnel. This compartment operates between 40-45°C. Similarly, the fifth compartment operates between 27-30°C. The manufacturing company of Citarshine recommends optimal working conditions of the drying air as Rh 25 to 30 per cent and a temperature below 60°C.

The fruits collected at the discharge end were rated sticky, similar conditions of fruits were also observed at waxing and grading unit of Chhauni Kalan. Such condition of fruits at the end of surface drying process is undesirable as the fruits have to undergo the size grading process immediately after the surface drying process. In practice the fruits are conditioned for about 30 minutes in crates under ambient conditions. The sticky condition of fruit surface may be attributed to the wax application rate, the temperature inside the tunnel or, the forward speed of the conveyor (Anonymous, 2001).

While rolling down from the waxing unit into the surface drying unit, it was observed that about one per cent fruits got damaged (squeezed) between the moving conveyer and the side support of the drying unit. The damage can be avoided by repositioning the chute between the waxing unit and the surface drying unit. In terms of temperature maintenance of drying air inside the drying tunnel, the performance of the surface drying unit can be rated as satisfactory. As far as surface condition of the fruits at the end of the drying process is concerned, the minor adjustments of the temperature controllers could lead to proper drying of the fruit surface.

Grading unit

The grading unit was evaluated for its performance as per the procedure given in section 4.5.4. The number of Kinnow fruits out of the sample of 50 fruits, graded in upper grade, exact grade and lower grade are shown in Table 4.2. It can be seen from Fig 4.9 that the grading efficiency is higher for bigger diameter fruits than the fruits with smaller diameter. The grading efficiency varies from 82% for smaller size fruit to 90% for bigger fruits.

Table 4.2 Grading of Kinnow

Size (mm)	No. of fruits	Correct size	Under/ over size	Efficiency (%)
46-55	50	41	9	82
56-65 50 43	7	86		
66-75 50 45	5	90		

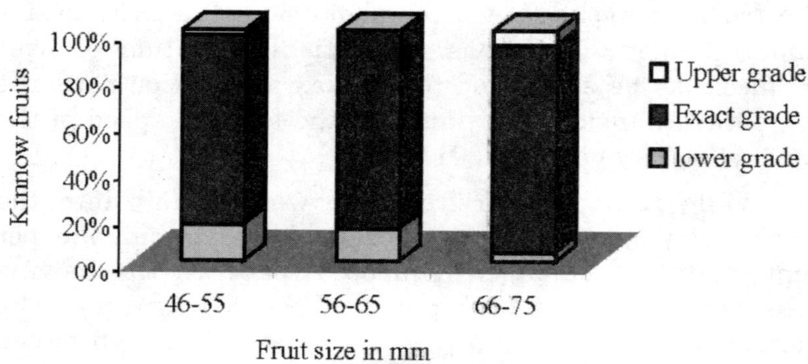

Fig 4.9. Number of Kinnow fruit into upper, exact or lower grade when graded with the newly developed mechanism

The performance evaluation of the developed prototype showed fairly satisfactory performance of the on-farm waxing grading prototype at 0.5 t /h capacity. There were a few incidences of damage occurring to the fruits during the operation which can be avoided by providing minor adjustments in the various sub-unit.

Chapter 5

QUALITY CHARACTERISTICS AND RECENT DEVELOPMENTS

One of the important objectives of the study was the comparison of the performance of prototype on the quality of the produce in respect of application of a suitable coating on the Kinnow fruit. There are several quality parameters which determine the final economic value of the product. The most important of these are: physical parameters – weight loss, firmness, juice content, and the decay; and chemical parameters – TSS, acidity, and the ratio of TSS to acidity. This chapter discusses the effects of different types of treatments/mechanical handling and subsequent storage of fruit on these quality characteristics; the non-destructives methods of internal and external quality determination are also briefly discussed.

POST HARVEST DISEASES

The black rot caused by *Alternaria citri,* blue and green mold rots caused by *Penicillium italicum* and *P. digitatum,* respectively, and stem-end rots caused by *Botryodiplodia theobramae, Diaporthe citri* and *Phomopsis* are serious diseases of citrus fruits all over the world. The sour rot caused by *Geotrichum candidum* and the rot caused by *Aspergillus niger* are also important in hot tropical areas particularly in India.

Fungicides such as Benomyl, Imazalil and etaconazol have been very effective in controlling rots. Recently biocontrol agents such as *Bacillus subtilis, Pichia guilliermondii* and *Trichoderma* spp. have been tried to control these rots. Safety from human consumption point of view besides economics, efficacy and feasibility of treatment are some of the points which are to be looked into for wide scale use and success of this technology (Maini and Ladaniya, 1999).

Dhatt *et al.* (1991) reported that uniform, healthy Kinnow mandarins were individually packed in 0.010 mm thick HDPE film bags (heat sealed, closed with rubber bands or twisted) and stored in wooden boxes at ambient temperature (7-32°C). Average decay was low after 4 week storage in wrapped fruits sealed using different methods; but after 12 week storage decay was 18.6% in sealed bags and 25% in fruits with twist bags.

Kumar *et al.* (1991) reported that uniform fruits of Kinnow mandarin were harvested at the full-ripe stage and treated with two per cent sesame oil emulsion or 1000 ppm $KNnO_4$ before being packed in wooden boxes lined with paper. Boxes of fruits for modified atmosphere storage were packed in polyethylene bags. All bags were stored for up to 84 days at room temperature. Decay was recorded at 14 day intervals and it was found that Sesame oil and $KMnO_4$ treatments reduced decay loss during the early stages of storage (upto 56 days) but were not effective in the later stages.

Dhatt *et al.* (1995) reported that Kinnow fruits were stored at ambient temperature (13-25°C) after giving treatments. Treated fruits were wrapped in HDPE film. Imazalil (500 ppm, 1000 ppm) and 2,4-D (250 ppm) were used as fungicides. Treated wrapped fruit emerged better and decay percentage in these fruits was minimum up to 80 days of storage as compared to untreated wrapped and unwrapped fruits. All the unwrapped fruits were shriveled with dried peel after 30 days and were totally shrunken at the end of storage.

Kaushal and Thakur (1996) reported that Kinnow mandarin were dipped for 5 minutes in one per cent bavistin (carbendazim) and then packed in sealed bags of 150 gauge polyethylene and stored for 8 weeks in evaporatively cooled chambers, constructed from low cost material available locally and under ambient conditions. Bags contained 1, 2 or 4 fruits or one kg of fruits. Spoilage due to fungal decay was greater in the fruits stored under ambient conditions and when more fruits were packed together. Fruits stored in cool chamber were more acceptable as they had a better appearance, texture, taste and flavour.

Ladaniya *et al.* (1998) reported that bavistin treated Nagpur mandarin were individually shrink wrapped in polyethylene and Cryovac (polyolefin material) films to extend their shelf-life at ambient (30-35°C; 25-35% Rh) and refrigerated (6-7°C; 90-95% Rh) conditions. The results revealed that bavistin (4000 ppm) treatment significantly minimized decay in shrink wrapped fruits under ambient and refrigerated storage conditions as compared to unwrapped fruits.

PHYSIOLOGICAL DISORDERS

Physiological disorders significantly influence the appearance and marketability of fruit sometimes causing considerable losses to growers and exporters. Many pre and post-harvest factors influence the occurrence of these disorders which are unpredictable. Important pre-harvest physiological disorders are puffiness, sunburn and granulation. The post-harvest disorders caused by harvesting injury; temperature, humidity, gas composition mismanagement and mechanical stress include rind staining, Kohansho, oleocellosis, Peteca (in lemons) stylar-end breakdown (in Tahiti limes) and chilling injury as watery breakdown or pitting all over the rind).

Tangerines and mandarins are very sensitive to rough handling and waxing, and they develop different symptoms such as rind staining and scattered collaps of flavedo due to necrosis of cell surrounding oil glands. Sometimes completely sunken brown coloured irregular patches develop on fruit surface.

The collar breakdown disorder in Nagpur mandarin was found to be due to bruises and injuries at coll during harvesting and handling. The dry (very low hurhidity) handling conditions, excessive brushing and high temperature further aggravated the disorder. It could be controlled by taking care of all these factors.

Recently in Fallglow tangerine pitting disorder characterised by irregular rind staining was found to be due to waxing with shellac and polyethylene based waxes and handling / storing at higher (>15.5°C) temperature. The disorder could be controlled when these waxes were avoided and fruit stored at low temperature (4.5°C) within hours after waxing.

The incidence of these physiological disorders can be greatly reduced by careful handling avoiding bruises and injuries and by proper management of temperature and humidity during storage, shipment and marketing.

Maximum research has been conducted in ameliorating chilling injury since sub-optimal lower temperatures (lower than the optimum cool storage temperature) were tried to reduce rotting and increase storage ability of the fruit. Moreover chilling injury occurrence is unpredictable and vary from cultivar to cultivar and 'depend on fruit maturity and growing conditions. Grapefruits, limes, lemons and pomelos are more chilling sensitive than oranges and mandarins. Temperature conditioning and treatment with TBZ and Imazalil have been reported to alleviate chilling in grapefruit and lemons.

Chilling injury in mandarins could be reduced by conditioning and intermittent warming. In Fortune

(Clementine X Dancy) mandarins chilling was less and storing qualities improved when fruits were dipped in hot (50~52 *DC*) water for 3 min and stored for 5 weeks with intermittent warming (at 10 *DC* for 3 days after every 4 days at 2 *DC*). These mandarins develop chilling below 10 *DC* if held for 4 weeks. In Olinda oranges, stored up to 25 weeks at 3 *DC,* the intermittent warming (3 weeks at 3 *DC* followed by 2 weeks at 15 *DC*) delayed onset of chilling injury and increased resistance to chilling by minimizing fatty acid unsaturation in lipids of flavedo.

POST HARVEST INSECT PROBLEM AND TREATMENTS

Fruit flies mainly Mediterranean fruit fly (*Ceratitis capitata* Wied), Caribbean fruit fly (*Anastrepha suspensa* Loew.) Mexican fruit fly (*Anastrepha ludens* Loew.) and West Indian fruit fly (*Anastrepha obliqua* Macquart) cause serious problem during postharvest handling and international trade of oranges and grapefruits in major citrus growing regions (including Florida and Mediterranean countries) of the world. Citrus fruits from these infested areas needs to be disinfested during export. Ethylene dibromide (EDB), ethylene oxide and methylene bromide (MEB) fumigation has been the practice for disinfestation since long but it has been banned for reasons of food safety and environmental protection.

Recently, several eco-friendly methods of fruit treatment including heat and cold treatments have been tried as post-harvest quarantine before or during the shipment. Since the low temperature required to kill larvae or eggs in the fruit were sub optimal temperatures (say 0-1 *DC* for 10-18 days) causing chilling injuries several techniques such as intermittent warming and coatings with sucrose fatty acid esters have been tried to alleviate chilling injury.

Considerable research has been done on use of vapour heat treatment to disinfest the fruit and find out fruit tolerance to this treatment. Moist forced air at 46 DC for upto 4 hrs was found promising for Valencia oranges where as moist hot air (46 DC, for 4.5 hrs or 50 DC for 24 hrs) resulted in acceptable flavour and market quality in Taxas N33 Navel orange. The standard condition for heat treatment (moist or dry) is 43.5 DC for eight hours.

The ionizing radiation have also been tried for disinfestation. The 300 Gy dose was lethal to all stages of *Ceratitis capitata* without altering chemical composition of Tarocco and Sanguinello orange cultivars grown in Italy. The advantages were deep and uniform penetration of energy and guaranteeing disinfestation even in the centre of the fruit without any undesirable effect. The treatment was effective in wrapped fruit also thereby preventing recontamination. In Florida, irradiation dose of 1.0 k Gy has been approved by FDA and found effective for control of Carribean fruit fly in grapefruit hipped to Japan and Europe. The vapour heat treatment (2 hrs at 38°C) reduced the severity and incidence of peel injury by 50% in fruit irradiated by 0.5-1.0 k Gy (Maini and Ladaniya, 1999).

PHYSICAL CHARACTERISTICS

Weight loss

The effect of weight loss of horticultural produce under different conditions of packaging and storage were studied by many researchers. Mahajan *et al.* (2002) reported the quality indicators of Kinnow fruit during the storage. The fruits were considered firm, slightly shriveled, and shriveled and unmarketable for weight loss of 0-2.5 %, › 2.5% - ‹ 5.5%, and › 5.5%, respectively. The Citrashine coated fruits stored under ambient conditions appeared slightly shriveled on 10th day of storage. The Kinnow fruits treated with Citrashine registered a slow and gradual increase in weight loss with the advancement of storage period and finally recorded 4.20% weight loss at the end of 2 moths of storage (Mahajan *et al.*, 2005).

The effectiveness of polymeric films in weight loss reduction was reported by many researchers (Ben-Yehoshua 1981, Purvis 1983, Jimenez *et al.,* 1983, Cohen *et al.,* 1990, Kaushal and Thakur 1996). For Kinnow fruits wrapped in polymeric films weight loss of 2-7 per cent was reported (Dhatt *et al.,* 1995, Kumar *et al.,* 1991, Dhatt *et al.,* 1991, Ladaniya *et al.,* 1997). Five fold reduction in weight loss for individually sealed packed citrus fruits of different varieties was reported Ben-Yehoshua (1978).

Firmness

The primary reason for the retention of firmness of shrink wrapped tomatoes was the decrease in water loss. Decrease in water loss is expected to be associated with significant reduction in the movement of water molecules (water stress) from the cell tomato flesh (Greg and Santi 1987). The unbound water, together with the bound water cells, maintains the firmness of fruits. The change in firmness of Red delicious apples during storage was related to the moisture loss from produce (Anzueto and Rizvi 1985). This caused the apples to become spongy and dry and thus more resistant to penetration by the pressure tester. Santi and Carl (1988) reported that fresh Redglobe peaches were treated with fungicide and shrink wrapped individually in polyolefin film. The wrapped and unwrapped fruits were stored at 1°C and 10°C, respectively, for period of 50 days. Samples of peaches were examined for textural qualities. The parameters such as hardness, brittleness, gumminess, elasticity and cohesiveness were measured throughout the entire storage period. All the parameters were observed to decrease during storage. A significant decrease in hardness between the wrapped and control fruit was observed. In all treatments the hardness of the wrapped fruits was lower than that of the control fruits. The change in hardness was nonlinear in the wrapped fruits. During the first week of storage, the hardness in these fruits reduced to 20-30% of the initial value and then it was maintained almost steady during the rest of storage period. In case of control fruits the hardness decreased gradually until the fourth week of storage and then started to increase again. This trend suggests that the control fruits, after losing 10-15% moisture, became firmer and tougher.

The effect of mechanical handling (vibrations) on fruits as reported by Prussia *et al.* (1989) revealed that in squashes the amount of bruising increased from 20% for the control to 70 per cent for the squash handled mechanically for one hour. Singh *et al.* (1992) evaluated the textural properties of tomatoes handled mechanically on vibration table under simulated transportation for different durations. The duration of vibration (mechanical handling) has a pronounced effect on the firmness. And the firmness decreased with longer duration of vibration. Similar results for mandarin fruits transported by rail (2600 km) were reported by Barua and Yamdagni (1996).

Juice content

The juice content of citrus fruits varies due to various factors such as variety, climate, cultural practices etc. Jawanda *et al.* (1973) obtained juice yield of 36.9 to 41.2 per cent by hand pressing of peeled Kinnow fruits. By using mechanical methods of juice extraction different workers have reported varied levels of juice yield from Kinnow fruit in percentage as : Bhullar *et al.* (1981) 46, Devkota *et al.* (1982) 46.08 to 55.41, Bal and Chohan (1983) 48.35 to 52.77, Mehta and Bajaj (1984) 50.7, Sandhu *et al.* (1980) 52.41, Sarmah *et al.* (1981) 57.7, Saini *et al.* (1985) 46.9 to 53.7, Pruthi *et al.* (1984) 55.8, Nagar (1993) 33.8 to 34.9, Chohan *et al.* (1966) 42, Singh and Singh (1967) 39, Sethi *et al.* (1980) 36.75, Lotha (1992) 42.84 to 49.07, Ranote *et al.* (1993) 56.5, while maximum juice recovery of 58.6 per cent was obtained from late season Kinnow by Ranote and Bains (1982).

Dhatt *et al.* (1991) reported that uniform, healthy Kinnow mandarins were individually packed in 0.010 mm thick HDPE film bags (heat sealed, closed with rubber bands or twisted) and stored in wooden boxes at ambient temperature (7-32°C). In general, fruits in sealed bags yielded better juice recovery than those with rubber bands or twisted HDPE bags.

Dhatt *et al.* (1995) reported that Kinnow fruits were stored at ambient temperature (13-25°C) after giving treatments. Treated fruits were wrapped in HDPE film. Imazalil (500 ppm, 1000 ppm) and 2,4-D (250 ppm) were used as fungicides. Juice percentage of wrapped fruits was almost maintained upto 60 days of storage and after that it started decreasing slowly. Maximum recovery of juice was recorded in fruits treated with imazalil (500 ppm) (47%) closely followed by imazalil (1000 ppm) after 80 days of storage. In unwrapped fruits, juice percentage was minimum i.e. 36% after 60 days of storage. The higher recovery of juice might be attributed to reduction in transpiration and respiration losses in HDPE wrapped fruits.

CHEMICAL CHARACTERISTICS

Total soluble solids (TSS)

Ting and Rouseff (1986) stated that total soluble solids of citrus juices are composed of 80% sugar, 10% of citric acid and its salts while the remainder constituting nitrogenous compounds and other minor soluble substances. TSS of Kinnow oranges grown under Punjab conditions have been recorded as 12, 11.8 and 13.5% from early, mid and late season crops harvested from Abohar area of Punjab (Ranote and Bains 1982). Saini *et al.* (1985) reported TSS of Kinnow juice to vary between 7 to 11° Brix. Sarmah *et al.* (1981) in their study on quality of Kinnow mandarin juice from fruits grown in sub-mountainous area of Hoshiarpur district reported TSS value of 13.4° Brix when harvested during last week of February. Jawanda *et al.* (1973) attributed TSS of Kinnow to fruit size varying from 9.7 – 12.2° Brix. They observed that smaller the size higher is the TSS. TSS value of 9.95 to 12.2° Brix has also been reported by Bal and Chohan (1983). Josan *et al.* (1983) observed that TSS ranged between 10.1 to 11.2° Brix in fruits harvested during second fortnight of December to first fortnight of February. Bhullar *et al.* (1981) reported TSS of 11.6° Brix in Kinnow juice. While lower TSS value of 6.9, 8.0 and 9.3 have been reported by Dhillon *et al.* (1977), Srivastava and Bopaiah (1978) and Singh *et al.* (1986), respectively.

Review of work conducted on composition of Kinnow juice indicated that TSS that TSS variations are attributable to fruit size, maturity and weather conditions. Smaller sized Kinnow fruit seems to be better suited for processing as far as TSS is concerned. Similarly, TSS content of fruit increases as the fruit ripens and it is maximum when the fruit is fully mature.

Dhatt *et al.* (1991) reported that uniform, healthy Kinnow mandarins were individually packed in 0.010 mm thick HDPE film bags (heat sealed, closed with rubber bands or twisted) and stored in wooden boxes at ambient temperature (7-32°C). Total soluble solids decreased in all the HDPE wrapped fruits.

Jain and Chauhan (1993) reported that Kinnow mandarin fruits packed in various packages, i.e. gunny bags, bamboo baskets, wooden boxes, wire bound boxes and corrugated paper boxes were transported by rail for 600 km to evaluate the effects of packages on their quality in transportation. The total soluble solids (TSS) of all the samples from different packages were compared during storage after transportation and it was found that their TSS concentration was not affected by the different packages.

Dhatt *et al.* (1995) reported that Kinnow fruits were stored at ambient temperature (13-25ºC) after giving treatments. Treated fruits were wrapped in HDPE film. Imazalil (500 ppm, 1000 ppm) and 2,4-D (250 ppm) were used as fungicides. Total soluble solids slightly decreased with storage in HDPE wrapped fruits. However, in some treatments, increase in TSS was observed after 60 days of storage. In warped control fruits, TSS increased up to 60 days of storage.

Acidity

Many organic acids have been found in citrus juice. Citric acid is by far the most predominant accounting for 85 to 95% of total acids in various cultivars (Ting and Rouseff 1986). L-malic acid also occurs in most cases and is the main acid found in the peel. The titrable acidity of the juices of most citrus fruits is due largely to citric acid. Clements (1964) analysed various citrus fruits and observed the presence of malic and citric acid to the extent of 0.08 to 0.2 and 0.56 to 0.98% respectively in orange juice. Sarmah *et al.* (1981) reported an acidity value of 0.65% in fully ripe Kinnow fruit juice. Ranote and Bains (1982) while studying juice from early, mid and late season Kinnow found the acidity value of 1.03, 0.7 and 0.79% respectively. Higher value of 1.47% for Kinnow has been reported by Bhullar *et al.* (1981) while Mehta and Bajaj (1984) reported an acidity value in the range of 0.4 to 1.0% for different component parts of Kinnow mandarin. The per cent acidity of juice from Kinnow was found out to be 0.59 by Srivastava and Bopaiah (1978), 1.15 by Sandhu and Bhatia (1985), 0.91 to 1.07 by Josan *et al.* (1983), 0.85 by Dhillon *et al.* (1977), 0.84 to 0.98 by Bal and Chohan (1983), 0.7 to 1.07 by Saini *et al.* (1985) and 0.6 to 1.12 by Jawanda *et al.* (1973). Highest acidity was recorded for early picking which declined with maturity (Bal and Chohan 1987, Josan *et al.,* 1983). The above studies indicate that Kinnow fruit contain good amount of acidity.

Dhatt *et al.* (1995) reported that Kinnow fruits were stored at ambient temperature (13-25°C) after giving treatments. Treated fruits were wrapped in HDPE film. Imazalil (500 ppm, 1000 ppm) and 2,4-D (250 ppm) were used as fungicides. Data reveals that acidity in all the treatments decreased remarkably with minimum value of 0.37% in untreated wrapped fruits after 80 days of storage. In unwrapped fruits, only a marginal decrease was observed where 0.55% acidity was recorded after 60 days of storage.

Kaushal and Thakur (1996) reported that Kinnow mandarins were dipped for 5 minutes in one bavistin (carbendazim) and then packed in sealed bags of 150 gauge polyethylene and stored for 8 weeks in evaporative cool chambers, constructed from low cost material available locally and under ambient conditions. Bags contained 1, 2 or 4 fruits or one kg of fruits. The combination of bavistin dips and polyethylene packing exhibited decrease in titrable acidity than fruits not sealed in bags, especially under cool chamber storage.

NON DESTRUCTIVE TECHNIQUES FOR QUALITY DETERMINATION

In order to develop an automated system for food quality evaluation, image processing techniques are often combined with mechanical and instrumental devices to replace human manipulative effort in the performance of a given process. In such a system, the image processing system is the centre, which controls the operation of the machinery. Li, Wang, and Gu (2002) developed an automated system for apple surface defect detection, which consisted of a feeding unit, an apple uniform spacing unit, a machine vision system, and a sorting conveyor. The apples were fed to the machine vision system for the defect inspection with the feeding and uniform spacing conveyors, and graded with the sorting unit. Mechanization is a remaining challenge in applying machine vision for food quality evaluation, especially for those easily bruised and marked when they are in contact with hard surface

The application potential of image processing techniques to the food industry has long been recognised (Tillet, 1990). The food industry ranks among the top ten industries using image processing techniques (Gunasekaran, 1996), which have been proven successful for the objective and non-destructive evaluation of several food products (Timmermans, 1998). The basic theory of computer vision technology for food quality assurance had been reviewed by Gunasekaran (1996). Image processing analysis generally consists of five steps (1) image acquisition operations to convert images into digital form; (2) pre-processing operations to obtain an improved image with the same dimensions as the original image; (3) image segmentation operations to partition a digital image into disjoint and non-overlapping regions; (4) object measurement operations to measure the characteristics of objects, such as size. shape, color and texture; and (5) classification operations to identify objects by classifying them into different groups.

Acquisition technology for internal structures

External attributes such as size, shape, colour, surface texture and external defects can be evaluated by ordinary means (e.g. CCD camera). However, internal structures are difficult to be detected with relatively simple and traditional imaging means, which cannot provide enough information for detecting internal defects, such as water-core, internal breakdown, and hollow heart. It is necessary to apply special image acquisition techniques for food quality evaluation. Ultrasound, MRI, CT, and ET technologies are potential solutions to the problem for inspecting internal attributes of food products. Applications of these three image acquisition techniques for food quality evaluation will be discussed separately below. Ultrasound is a new technology that offers a route to acquire internal images for use in food quality evaluation. There are two modes for ultrasonic imaging of biological tissues. One is A-mode (amplitude modulation) and the other is B-mode (brightness modulation). While A-mode is one-dimensional and is limited to measuring depth of tissue, B-mode allows for characterization of tissue with different densities. Real-time ultrasound (RTU) is a specialized version of B-mode ultrasound and has the ability to produce images of moving objects almost instantaneously. As a cost-effective and reliable method, the majority of applications of ultrasonic imaging in food quality evaluation can be found in the meat industry. Ultrasonic image analysis has been performed for measurement of fat thickness estimation of yield, and assessment of quality of meat.

MRI is an imaging technique primarily used to obtain high quality images of the inside of the object in two or three dimensions. Numerous publications in the literature have indicated that MRI can be used to noninvasively evaluate important quality attributes of food products. Being an analytical tool, MRI is based on the absorption and emission of energy in the radio frequency range of the electromagnetic spectrum. Generation of magnetic resonance images can be controlled by the radio frequency pulse sequence used for excitation of the nuclear spins. Selection of the optimal pulse sequence parameters can result in enhanced images, with good contrast between a region of interest and its surrounding area. In some instances, MRI research was conducted to measure moisture and its migration in food systems; other researches demonstrated the potential of using MRI to investigate physical or biological properties of food products; still some other work sought to estimate the yield.

CT is a non-destructive technique for capturing food images, which allows visualization of the internal features of an object. Based on the differential that exists between the rates at which the tissues in the object attenuate X-rays, the image of a thin slice of the object can be obtained using a movable X-ray source and detector assembly to accumulate data. A three-dimensional representation of the object can be made by stacking several slices of the scanned object and stored digitally and used later for evaluation. It is a proven efficient method for evaluating a cross-section of an object and has received extensive applications in the food industry. This method was used wide spreadly for many food products, including fishery, fruits, meat, and vegetables.

In addition, CT can be used for detection of foreign materials in foods. Ogawa *et al.* (1998) used a medical X-ray CT scanner as a non-destructive inspection method for the detection of selected non-metallic materials embedded in various fluids and food materials. The limitation of using X-ray linear absorption coefficients from typical CT systems is to detect certain combination of foreign materials and foodstuffs.

ET techniques for food image acquisition are becoming increasingly popular, which mainly consist of the following three sensing modalities: electrical resistant (impedance) tomography (ERT or EIT), electrical capacitance tomography (ECT), and electromagnetic inductance tomography (EMT). These methods exploit differences in the electrical properties of different materials to produce slice-images and they are relatively fast, non-destructive, and low-cost imagers. They are based on the measurement and computerised analysis of electrical resistively, capacitances, and inductance changes of an object for ERT (ElT), ECT, and EMT, respectively. There are wide applications using ET techniques to improve the design and operation of equipment and ensure consistent and high quality of food product. Primrose and Bolton (200 I) reported that ER T could be applied to determine presence of non-conducting foreign objects, in-line concentration of particular ingredients, degree of mixed ness in static and batch mixers, bubbles and structure in foods. EIT has been used for food temperature measurement (Nott & Hall, 1999). Potential applications for EMT techniques are where the material distribution can be characterised by either a high electrical conductivity or ferro-/ferri-magnetic behaviour, such as food inspection (Peyton *et at 1999)* and measurement of the moisture content of grains (Abdullah, Guan, and Karim, 2004). A simple ECT sensor has been developed to monitor the number of fruit pieces entering a canning machine (IFR, 2003). The drawback of ET is its relatively low spatial resolution.

In order to assure the internal quality with respect to desired total soluble solids (sugars) and acidity in Satsuma mandarins, a non-destructive on-line measurement technique using Near Infra Red Spectroscopy in the transmittance mode has been developed for use in packing houses.

On the basis of relationship between the emission of volatile substances and the loss of quality during transportation, the sensors were developed to monitor the quality of citrus fruits.

Some of the most frequent instances in which quality deteriorates during transportation are (1) Decay caused by *Penicillium digitatum, Botrytis cinerea and Geotrichum candidum* (2) Damage caused by bad handling the volatiles emitted by disease pathogens were studied in order to develop the sensors.

Two characteristic emissions related to decay caused by *Penicillium digitatum* were of ethanol and ethyl acetate. No significant emissions of Limonene were noted. Only a small amount of limonene was noted in case of *Botrytis cinerea*. A strong emission of limonene was noticed in case of *Geotrichum Candidum*. Due to physical damage presence of significant amounts of limonene was observed with no other characteristic emissions.

The development of sensors for the limonene (due to recent damage/decay), alcohols (due to decay/ageing), aldehydes (due to ageing) and ethyl acetate (due to decay) were reported to be possible.

Two different kinds of sensors were reported.

1. Chemical sensors

(i) Tin dioxide gas sensors (limonene, alchols and ethyl acetate) (ii) Polypirrol (ethyl acetate and alchols)

2. Biochemical sensors

(i) Enzymes /polypirrol sensor (aldehydes)

The prototypes called Fruit box and Sniffers have been developed. The Fruit box were made up of sensors (for Temperature, Humidity and volatiles), signal conditioning circuits and a microcontroller, to record and store the data output from the sensors, thus enabling continuous monitoring of fruit conditions during a shipment. The importer at the destination will be able to interface the microcontroller card to its PC and through a specially developed software module gain access to the stored data which will be compared with the data provided by the packing house. Therefore, any physical damage or quality degradation in fruit, caused during transport will be identified. The fruit Sniffers were reported to have utility to check individual pallets and boxes for damaged and mouldy fruit (Maini and Ladaniya, 1999).

Chapter 6

BIBLIOGRAPHY

A.S.T.M. (1980) *Standard methods of test for water transmission of thick materials*. Am. Soc. Test Master. **18**: 124-131. In: Miller W M and Grierson W (1983) Properties data for coatings on citrus. *Trans ASAE* **26**: 1884-1887.

A.O.A.C. (1989) Official and Tentative Methods of Analytical Chemists. Washington D. C., U S, 14th edition.

Abdullah M Z, Guan L C, Lim K C and Karim A A (2004) The applications of computer vision system and tomographic radar imaging for assessing physical properties of food. *Journal of food Engg*. 61(1), 125-135.

Akimoto K and Maezawa (1997) A new method for estimating respiration rate of fruits and vegetables in modified atmosphere packaging. *J Japanese Soc of Agril Machinery* **59**: 109-116 (Orizinal not seen. Abst from CAB Abstracts 1996-98/07).

Alam M S (1998) *Effect of diffusion barriers on juice recovery and shelf-life of Kinnow*. M.Tech. Thesis, Punjab Agricultural University, Ludhiana, India.

Allhouse G W and Stephenson K Q (1969) Development of a handling and sorting system for certain fruits and vegetables. *Trans ASAE* **12**: 290-292.

Andres C (1984) Natural edible coating has excellent moisture and gas barrier properties. *Food Processing* **45(13)**: 48-49.

Anonymous (1992) CSIRO Division of Material Science and Technology. Australian challenges (operations) Pvt Ltd. (www.netaddress.com).

Anonymous (2001) *Horticulture - Citrus.* South Australian Research and Development Institute (www.sardi.sa.gov. au).

Anonymous (2003) *Statistical Abstract of Punjab.* pp 191-193. Economic & Statistical Organization (Punjab), Chandigarh.

Anonymous (2004) *Packingline Operations.* Texas Agricultural Extension Service (www.extension-horticulture.tamu.edu).

Anonymous (2006) *Horticulture Information Services.* National Horticulture Board, Ministry of Agriculture, Govt of India.

Anzueto C R and Rizvi S H (1985) Individual packaging of apples for shelf life extension. *J Food Sci* **50**: 897-900.

Arvanitoyannis I and Biliaderis C G (1998) Physical properties of polyol-plasticized edible films made from sodium caseinate and soluble starch blends. *Food Chem.* 62:333-342.

Arvanitoyannis I and Gorris L G M (1999) Edible and biodegradable polymeric materials for food packaging or coating. In: F A R Oliveira, J C Oliveira, M E Hendrickx, D Knorr and L G M Gorris (eds). *Processing foods – Quality Optimization and Process Assessment.* Boca Raton, FL CRC Press pp. 357-370.

Avenainen R, Myllarinen P and Poutanen K (1997) Prospects of using edible and biodegradable protective films for foods. *Europ. Food Drink Rev.*, Summer: 73, 75, 77-80.

Baker R A, Baldwin E A and Nisperos-Carriedo M O. *Lancaster, P A*: Techomic Publishing Co. Inc. pp. 89-104.

Bal J S and Chohan G S (1983) A study on distribution and quality of different sized Kinnow mandarin oranges harvested at optimum maturity. *J Food Sci Tech* **20**: 321-322.

Baldwin E A (1994) Edible coatings for fresh fruits and vegetables: past, present and future. In: Edible Coatings and Films to improve Food Quality, eds. J.M. Krochta E A, Baker R A, Baldwin E A and Nisperos-Carriedo M O. Lancaster, P A: Techomic Publishing Co. Inc. pp. 25-64.

Baldwin E A (1999) Surface treatments and edible coatings in food preservation. In: M S Rahman (ed) Handbook of Food Preservation. New York: Marcel Dekker, pp. 577-609.

Baldwin E A, Nisperos-Carriedo M O and Baker R A (1995) Edible coatings for lightly processed fruits and vegetables. *Hort. Sci.*, 30:35-38.

Baldwin E A, Nisperos-Carriedo M O and Baker R A (1995) Use of edible coatings to preserve quality of lightly (and slightly) processed products. *Crit. Rev. Food Sci. Nutr.* 35:509-524.

Ballantyne A, Stack R and Selman J D (1988) Modified atmosphere packaging of shredded lettuce. *Int J Food Tech* **23**: 267-274.

Banks H J (1981) Effects of controlled atmosphere storage on grain quality: a review. *Food Tech in Australia* **33**: 337-340.

Barua P C, Yamdagni R and Kaushik R A (1993) Effect of stages of maturity and packages on physical characteristics of fruits of mandarin (*Citrus reticulate* Blanco) during transportation and storage. *Haryana J Hort Sci* **22(2)**: 135-139.

Ben-Yehoshua S (1978) Delaying deterioration of individual citrus fruit by seal packaging in film of high density polyethylene – General effects. *In: Cary P R (ed), Proceedings of the Int Soc of Citriculture, Sydney, Australia, pp 110 -115.*

Ben-Yehoshua S (1985) Individual seal-packaging of fruits and vegetables in plastic film- A new postharvest technique. *Hort Sci* **20**:32-37.

Ben-Yehoshua S, Kobiler I and Shapiro B (1981) Effects of cooling versus seal packaging with high density polyethylene on keeping qualities of various citrus cultivars. *J Amer Soc Hort Sci* **106 (5)**: 536-540.

Bhowmik S R and Hulbert G J (1989) Effect of individual shrink wrapping of on shelf life eating quality and respiration rate of tomatoes. *Lebensmittel Wissenschaftund-Tschnologie* **22 (3)**: 119-123.

Bhowmik S R and Sebris S (1988) Quality and shelf life of individually shrink wrapped peaches. *J Food Sci* **53 (2)**: 519-522.

Bhullar J S and Farnaham H L (1980) Studies on the ripening and storage behaviour of safeda guava. *Indian Food Packer* **34**: 5-7.

Bhullar J S, Dhillon B S and Randhawa J S (1985) Effect of wrappers on the storage of Kinnow mandarin. *J Res (PAU)* **22**: 663-666.

Bhullar J S, Khokher U and Agnihotari R P (1981) Storage behaviour of Kinnow mandarin. *Pb Hort J* **21**: 27-30.

Bohling H and Hellickson M L (1998) Physiological responses of red delicious apples during controlled atmosphere storage. *Trans. ASAE* **41**: 675-678.

Bowman E (1975) *Efficiency in Manually Grading Citrus Fruit.* Packinghouse Newsletter No 73. Univ of Florida, AREC, Lake Alfred. In: Wadowaski W F, Nagy S and Grierson W (ed) (1986) *Fresh Citrus Fruits.* AVI AVI Publishing Co., Westport, CT.

Bureau of Indian Standards (1974) *Guide for Storage of Citrus Fruits*. IS: 7192.

Cabezas A, Richardson D G and Gorny J R (1997) Modified atmosphere packaging of broccoli florets: Effects of temperature and package type, *Proceedings of 7th Int Ccontrolled Atmosphere research Conference,* UC Devis, California, USA. (Orizinal not seen. Abst from CAB Abstracts 1996-98/07).

Callegarin F, Quezada Gallo J A, Debeaufort F And Voiley A (1997) Lipids and biopackaging. *JAOCS* 74:1183-1192.

Cantwell M I (1997) *Postharvest Handling of Specialty Vegetables*. Dept. of Vegetables Crops, Univ. California, USA.

Chien P J, Shen F and Lin H R (2006) Coating citrus (Murcott tangor) fruit with low molecular weight chitosan increases postharvest quality and shelf life. (Abstr available on sciencedirect.com).

Chohan G S, Nuriyal J P and Bakhsi J C (1966) Stidies on the performance of some promising citrus varieties at Abohar (Punjab). *PbHort J* **6**:56-61.

Christie G B Y, Macdiarmid J I, Schliephake K and Tomkin R B (1995) Determination of film requirements and respiratory behaviour of fresh produce in modified atmosphere packaging. *Postharvest – Biology and Technology* **6**: 41-54.

Clarke R, De-Moor C P and Gorny J R (1997) The future in film technology: A renewable packaging system for fresh produce. *Proceedings of 7th Int Ccontrolled Atmosphere research*

Conference, UC Devis, California, USA. (Orizinal not seen. Abst from CAB Abstracts 1996-98/07).

Clements R L (1964) Organic acids in citrus fruits: Varietal differences. *J Food Sci* **29**: 276-280.

Cohen E, Lurie S, ShapiroB, Ben-Yehoshua S, Shalom Y and Rosenberg I (1990) Prolonged storage of lemons using individual seal packaging. *J Amer Soc for Hort Sci* **115(2)**: 251-255.

Cuq B, Gontard N and Guibert S (1994) *Edible films and coatings and active layers*. In: M L Rooney (ed) *Active Food Packaging*. London: Blacke Academic and Professional pp. 111-142.

Daniels J A , Krishnamurthi R and Rizvi S S H (1985) A review of effects of CO_2 on microbial growth and food quality. *J Food Protection* **48**: 532-537.

Daniels R (1973) *Edible Coatings and Soluble packaging park*, Ridge NJ: Noyes Data Corp.

Das R C and Dash J (1967) The effect of wax emulsion 2,4-D and 2,4,5-T on the storage behaviour of Musamabi fruits. *Sub-tropical and Tropical Hort,* New Delhi: 104-107.

Dawson a J and Eckert J W (1977) Problems of decay control in marketing citrus fruits: Strategy and solutions, California and Arizona. *Proc Int Soc Citriculture.*

De Mulder-Johston C (1999) Thermal *Analysis of and Oil Migration through Films from, Why Protein Isolate*. Ph.D. Thesis, University of California, Davis. USA.

Deak T, Heaton E K, Hung Y C and Beauchat L R (1987) Extending the shelf life of fresh sweet corn by shrink wrapping, refrigeration and irradiation. *J Food Sci* **52 (6)**: 1625-1631.

Debeaufort F, Quezada-Gallo J A and Voiley A (1998) Edible films and coatings: tomorrow's packaging: A review:" *Crit. Rev. Food Sci.*, 38:299-313.

Devkota R P, Grewal S S and Dhatt A S (1982) Maturity determination of Kinnow mandarins. *Pb Hort J* **22(3-4)**: 131-135.

Dhatt A S, Randhawa J S and Singh S N (1991) Effects of individual seal packaging in high density polyethylene

(HDPE) film on storage life and quality of Kinnow. *J Plant Sci Res* **7**: 84-85.

Dhatt A S, Randhawa J S, Kaur H, Singh G and Sharma R (1995) Seal packaged storage of Kinnow fruits under ambient conditions. *Phala Samaskarana – 95, Proceedings of National Seminar on Postharvest Tech of Fruits and Vegetables held at UAS Bangalore.*

Dihllon B S, Bains P S and Randhawa J S (1977) Studies on the storage of Kinnow mandarin. *J Res (PAU)* **14**: 434-438.

Donhowe I G and Fennema O (1993) Water vapour and oxygen permeability of wax films. *J Amer Oil Chemists Soc* **70(9)**: 867-873 (Original not seen. Abstr in FSTA, Entry No 199401, 1993).

Echeverria E and Valich J (1989) Enzymes of sugar and acid metabolism in stored Valencia oranges. *J Amer Soc Hort Sci* **114**: 445-449.

Exama A, Arul J, Lencki R W, Lee L Z and Toupin C (1993) Suitability of plastic films for modified atmosphere packaging of fruits and vegetables. *J Food Sci.* **58**: 1365-1370

Fallstad M N (1966) Mycelial growth and sporulation of *Alternaria tenius,* Botrytis cinerea cladosporium herbarium and Rhizopus stalonifer in low O_2 atmosphere. *Phytopathology* **56**: 1089-1090.

Farooqui W A, Salih A and Zain-Ul A (1988) Effect of wax coating on the physiological and biochemical aspects of Kinnow fruit. *Pakistan J Sci Indstr Res* **31(2)**: 142-145.

Farooqui W A, Shankar G A, Malik M A and Ahmed M S (1981) Studies on the inhibition of *Alternaria citri* in stored citrus fruits. *Proc Florida State Hort Soc* **94**: 279-280.

Felsenstein G and Manor G (1973) A feasibility study into the development of an improved photoelectric device for sorting of citrus fruit for surface defects. *Trans ASAE* **16**: 1006-1009.

Gaffney J J (1973) *Potentials for photoelectric grading equipment in the Florida citrus industry*. Citrus Eng Conf XIX, Fla. In: Wadowaski W F, Nagy S and Grierson W (ed) (1986) *Fresh Citrus Fruits.* AVI AVI Publishing Co., Westport, CT.

Garcia H M D and Galvan E J G (2003) Development and construction of a machine for waxing fruits and horticultural products. *AMA*. **34**: 43-45

Geeson J D (1989) Modified atmosphere packaging of fruits and vegetables. *Acta Hort*. **258**: 143-150.

Geeson J D, Brown K M, Maddison K, Shepherd J and Guaraldi F (1985) Modified atmosphere packaging to extend the shelf life of tomatoes. *J Food Tech*. **20**: 339-349.

Gennadios A, and Weller C L (1990) Edible films and coatings from wheat and corn proteins. *Food Technol*. 44(10)63-69.

Gennadios A, Hanna M A and Kurth L B (1997) Application of edible coatings on meats, poultry and seafoods: A review. *Lebensm – Wiss u Technol*., 30:337-350.

Goodman H C and Hamann D D (1971) A machine to field size sweet potatoes. *Trans ASAE* **16**: 3-6.

Gorny J (1997) *Postharvest Handling of Horticultural Produce*. Paper presented at PAU, Ludhiana, India.

Greg J H and Santi R B (1987) Quality of fungicide treated and individually shrink wrapped tomatoes. *J Food Sci* **52(5)**: 1293-1297.

Grierson W (1957) Preliminary studies on cooling Florida oranges prior to packing. *Proc Florida State Hort Soc* **70**: 264-272.

Grierson W (1971) Trash elimination. Packinghouse Newsletter No 39. Univ of Florida, AREC, Lake Alfred. In: Wadowaski W F, Nagy S and Grierson W (ed) (1986) *Fresh Citrus Fruits*. AVI AVI Publishing Co., Westport, CT.

Grierson W and Wardowaski W (1973) Developmenmt of mechanization programs for harvesting and packing citrus for the fresh fruit markets. *Proc Int Soc Citriculture*.

Guilbert S (1986) Technology and application of edible protective films. In: M Mathlouthi (ed) Food packaging and preservation: Theory and practice, New York Elsevier Applier Science Publishers pp. 371-394.

Guilbert S, Cuq and Gontard N (1997) Recent innovation in edible and/or biodegradable packaging materials. *Food Additivies Contam*., 14:741-751.

Gunasekaran S (1996) Computer vision technology for food quality assurance. *Trends in Food Science & Technology*, 7, 245-256.

Hagenmaier R D and Baker R A (1993) Cleaning affect shrinkage rate of citrus fruits. *Hort Sci* **28(8)**: 824-825 (Original not seen. Abstr in FSTA, Entry No 199404).

Hagenmaier R D and Baker R A (1994) Wax microemusions and emulsions as citrus coatings. *J Agric Food Chem* **42(4)**: 899-902 (Original not seen. Abstr in FSTA, Entry No 199409).

Hagenmaier R D and Baker R A (1995) Some wax based edible coating formulations for fruit and other foods. *IFT annual meeting* pp 82 (Original not seen. Abstr in FSTA, Entry No 199511)

Hagenmaier R D and Baker R A (1996) Edible coatings from candelilla wax microemulsions. *J Food Sci* **61(3)**: 562-565.

Hagenmaier R D and Shaw P E (1992) Gas permeability of fruit coating waxes. *Amer Soc Hort Sci* **117(1)**: 105-109.

Hardenburg R E (1971) Effect of in-package environment on keeping quality of fruits and vegetables. *Hort Sci*. **6**: 198-201.

Hardenburg R E, Watada A E and Wang C Y (1986) The commercial storage of fruits, vegetables, and florist and nursery stocks. USDA Agril Handbook **66**: 75-92.

Hayakawa K, Heing Y S and Gilbert S G (1975) Formulae for predicting gas exchange of fresh produce in polymeric films package. *J Food Sci* **40**: 186-191.

Heaton E K, Dobson J W, Lane R P and Beauchat L R (1990) Evaluation of shrink wrap packaging for maintaining quality of apples. *J Food Protection* **53(7)**: 598-599.

Heft M E Jr and Wiant D E (1962) Lighting apple packing areas. *Illum Eng* **57(6)**: 5. (Original not seen, In: Wadowaski W F, Nagy S and Grierson W (ed) (1986) *Fresh Citrus Fruits.* AVI Publishing Co., Westport, CT).

Heing Y S and Gilbert S G (1975) Computer anlysis of the variables affecting respiration and quality of produce packaged in polymeric films. *J Food Sci* **40**: 1033-1035.

Hocking A D (1990) Response of fungi to modified atmosphere. *CSIRO Food Res Quatr* **48(3-4)**: 56-65.

Hood C E, McHugh C M, Sims E T Jr, Garrett T R and Williamson (1981) Orchard fruit handling system. *Trans ASAE* **27**: 20-22.

Howard L R and Dewi T (1995) Sensory, mocrobiological and chemical quality of mini-peeled carrots as affected by edible coating treatment. *J Food Sci* **60(1)**: 142-144.

Hughes P A, Thompson A K and Seymour G B (1981) Storage of capsicum under controlled atmosphere, modified atmosphere and hypobaric conditions. *J Hort Sci* **56**: 261-265.

Hulbert G J and Bhowmik S R (1987) Quality of fungicide treated and individually shrink wrapped tomatoes. *J Food Sci* **52(5)**: 1293-1297.

Hunter D L, Kafer F and Meyer C H (1958) *Apple sorting: Methods and equipments*. Marketing Res Rept 230. USDA Agricultural Marketing Services..

IFR (2003, 10 August) Electrical tomography and overview. Techniques in institute of food Research. Colney, Norwich NR4 7UA, UK: Norwich Research Park. www.ifr.bbsrc.ac.uk/Materials/sensors/tomography.

Jain P K and Chauhan K S (1993) Evaluation of various packages for transportation of Kinnow mandarin fruits by rail. *Haryana J Hort Sci* **22(1)**: 24-30.

Jawanda J S, Arora J S and Sharma J N (1973) Fruit quality and maturity studies of Kinnow mandarin at Abohar. *Pb Hort J* **13(1)**: 3-12.

Jawanda J S, Singh R and Vij V K (1978) Studies on extending post harvest shelf life of Kinnow mandarin. *Pb Hort J* **18(3-4)**: 149-153.

Jimenez C M, martinez J M and Cuquerella J (1983) Plastic individual seal packaging of Spanish citrus fruit. *In Refrigeration in the service of man. XVIth Int Congress of Refrigeration, Paris 3:251-257.*

Josan J S, Chauhan G S and Bal J S (1983) Kinnow fruit quality as affected by locations on the tree. *Indian J Hort* **1**: 40-42.

Kader A A (1980) Prevention of ripening in fruits by use of controlled atmosphere. *Food Tech* **34(3)**: 51-54.

Kaplan H J (1986). Washing waxing and colour adding. In: Wadowaski W F, Nagy S and Grierson W (ed) (1986) *Fresh*

Citrus Fruits. pp 379-395. AVI AVI Publishing Co., Westport, CT.

Kaplan H J and Petrie J F (1982) *Citrus.* In Decco Instructions Book. Pennwalt Corporation, Monrovia, CA.

Kaushal B B L and Thakur K S (1996) Influence of ambient and evaporative cool storage conditions on the quality of polyethylene packed Kinnow fruit. *J Advances in Hort Sci* **10(4)**: 179-184.

Kester J J and Fennema O R (1986) Edible films and coatings: A review. *Food Technol.* 40(12):47-59.

Kitinoja L and Gorney J (1998) *Postharvest Technology for Fruit and Vegetable Produce Marketers.* A joint publication of U C postharvest programe and Punjab Horticultural Postharvest Technology Center, USAID/ACE Project.

Koelsch C (1994) Edible water vapor barriers: Properties and promise. *Trends Food Sci. Technol.* 5:76-81.

Krochta J M (1992) *Control of mass transfer in foods with edible coatings and films.* In: R P Singh and M A Wirakartakasumah (eds) *Advances in Food Engineering*, Boca Raton. FL: CRC Press. Pp.517-538.

Krochta J M (1997) Edible composite moisture barrier films. In: B Blakistone (ed) (1996) *Packaging Yearbook*, Washington, DC: National Food Processors Association, pp.38-54.

Krochta J M (1997) Edible Protein Films and Coatings. In: S Damodaran and A Paraf (eds) *Food Proteins and Their Applications.* New York: Marcel Dekker, pp. 529-549.

Krochta J M (1997) *Films edible.* In: A L Brody and K S Marsh (eds) *The Wiley Encyclopedia of Packaging Technology*, 2nd Edition, New York: John Wiely & Sons, pp. 397-401.

Krochta J M and De Mulder-Johnston C (1997) Edible and biodegradable polymer films: Challenges and opportunities. *Food Technol.*, 51(2): 61-74.

Krochta J M, Baldwin E A and Nisperos-Carriedo M O (1994) *Edible coatings and films to improve food quality lancaster pa*: Technomic Publishing Company, Inc.

Kubo Y, Inaba A and Nakamura R (1990) Respiration and C_2H_4 production in various harvested crops held in CO_2 enriched atmosphere. *J Amer Sco Hort Sci* **115**: 975-978.

Kumar J and Sharma R K, Ran S and Singh R (1991) Effect of modified atmosphere storage on shelf life of Kinnow. *Haryana J Hort Sci* **20(3-4)**: 156-160.

Kumar R (1999) *Effect of edible coatings on shelf-life of and quality of okra*. M.Tech. thesis, Punjab Agricultural University, Ludhiana, India.

Ladaniya M S (2001) Post-harvest technology of fresh citrus fruits. In: Singh S and Naqvi S A M H (ed) *Citrus*. pp 459-473. International Book Distributing Co., Lucknow, India.

Ladaniya M S, Sonkar R K and Das H C (1998) Evaluation of heat shrinkable film wrapping of Nagpur mandarin (Citrus reticulta Blanco) for storage. *J Food Sci Tech* **34(4)**: 324-327.

Ladaniya M S, Sonkar R K and Singh S (1999) Shelf life of Nagpur mandarin fruits wrapped with heat shrinkable and stretch cling film under ambient conditions. *Proc Int Symposium on Citriculture.* NRC for Citrus, Nagpur, Maharashtra, India.

Langston D and Pfunter S (1986) Shrink wrapped grape fruit – an advantage for retailers. *Citrograph* **71(3)**: 27-31.

Lerdthanangkul S and Krochta J M (1996) Edible coating effects on postharvest quality of green bell peppers. *J Food Sci* **61(1)**: 176-179.

Li Q Z, Wang M H and Gu W K (2002) Computer vision based system for apple surface defect detection. *Computers and Electronics in Agriculture*, 36(2-3), 215-223.

Lillehog E B, Milburn M S and Ciegler A (1972) Control of *Pencillium martensii* development and penicillic acid production by atmospheric gases and temperature. *Applied Microbiology* **24**: 198-201.

Lindstrom T R, Morimoto K and Cante C J (1992) *Edible films and coatings*. In: Y H Hui (ed) *Encyclopedia of Food Science and Technology*. New York John Wiley & Sons pp. 659-663.

Long W G (ed) (1964) *Better handling of Florida's fresh citrus fruit*. Florida Agric Exp Stn Bull. 681.

Lotha R E (1992) *Studies on processing and storage of Kinnow mandarin juice*. Ph. D. dissertation. IARI, New Delhi, India.

Mahajan B V C, Dhatt A S and Randhawa J S (2003) Waxing preserves freshness and quality of Kinnow. *Progressive Farming* **39**: 24-25

Mahajan B V C, Dhatt A S and Rattan G S (2002) Evaluation of various wax formulations on the postharvest characteristics of Kinnow. *Indian J Citriculture* **1(2)**: 185-188.

Mahajan B V C, Dhatt A S and Sandhu K S (2005) Effects of different post-harvest treatments on the storage life of Kinnow. *J Food Sci Tech* **42(4)**: 296-299.

Maini S B and Ladniya M S (1999) Recent trends in postharvest management of citrus. *Proc Int Symposium on Citriculture.* NRC for Citrus, Nagpur, Maharashtra, India .

Mann S S (1978) Harvesting and storage of Kinnow fruits. *Pb Hort J.* **18 (3-4)**: 154-155

Mann S S and Sandhanwalia H S (1977) Effect of waxing and polyethylene packs on low temperature storage of Kinnow mandarin. *Paper presented at the Symposium on Citriculture held at Bangalore.*

Mannapperuma J D and Singh R P (1987) A computer aided model for gas exchange in fruits and vegetables in polymeric packages. *ASAE Paper No. 87-6526.* St. Joseph, Michigon, USA.

Marcellin P (1974) Conservation des fruits and legumes en atmosphere controlee a l' aide de membrane de polymers. *Revue generale du fraid* **3**: 217-236 (Original not seen. Cited by Exama A, Arul J, Lencki R W, Lee L Z and Toupin C, 1993, *J Food Sci.* **58**: 1365-1370).

Mc Guire R G (2002) Application of *Candida - guilliermondii* incommercial citrus coatings for bio control of *Pecnillium digitatum* on grapefruits. USDA ARS (Abstr available on sciencedirect.com).

Mchugh T H and Krochta J M (1994) Milk protein based edible films and coatings. *Food Technol* 48(1): 97-103.

Mehta U and Bajaj S (1984) Physico-chemical characteristics of Kinnow, Blood Red and Villafranca. *Pb Hort J* **24(1-4)**: 40-45.

Miller W M (1978) Surface moisture drying analysis of citrus fruits. *Trans ASAE* **21**: 1237-1241.

Miller W M and Grierson W (1983) Properties data for coatings on citrus. *Trans ASAE* **26**: 1884-1887

Mitra S K (ed) (2005) Postharvest Physiology and Storage of Tropical and Subtropical Fruits. pp 21-40. CABI Publishing, New Delhi, India .

Moleyar V and Narsimham (1994) Modified atmosphere packaging of vegetables : An appraisal. *J Food Sci Tech* **31**: 267-278.

Moyls A L (1998) Whole-bag (water) method for determining water vapour transmission rate of polyethylene films. *Trans ASAE* **41**: 1447-1451.

Moyls A L, Mc Kenzie D L, Hocking R P, Toivone P M A, Delaquis P, Girard B and Mazza G (1998) Variabilityy in O_2, CO_2, and H_2O transmission rates among commercial polyethylene films for modified atmosphere packaging. *Trans ASAE* **41**: 1414-1446.

Myllarinen P, Rantamaki P, Latva K and Ahvenainen R (1997) Minimization of Food Packaging Using Active Edible Coatings Possibilities and Challenges Finland: VIT Information Service.

Nagar P K (1993) Effect of different harvesting periods on shelf-life and quality of Kinnow fruits. *J Food Sci Tech.* **30(1)**: 44-45.

Nisperos C and Baldwin E A and Shaw P E (1991) Development of an edible coating for extending postharvest life of selected fruits and vegetables. *Proc Florida State Hort Soc* **104**: 122-125.

Nisperos-Carriedo M O (1994) Edible coatings and films based on polysaecharides. In: J M Krochta, E A Baldwin and M O Nisperos-Carriedo (eds) Edible coatings and films to improve food quality:

Nott K P and Hall L D (1999) Advances in temperature validation of foods. *Trends in Food Science & Technology*, 10, 366-374.

Nowak J and Rudnicki R M (1984) Cold storage of cut gladiolus spikes. *Rosliny Ozdobne, Ser B* **9**: 67-72 Skierniewice, Poland.

Noznick P P and Bundus R H 1967 U.S. Patent 3.314.800.

Nussinovitch A (1998) Hydrocolloid coating of foods: A Review, Leatherhead food RA. *Food Industry Journal*, 1:174-188.

O'Connor, Skarshewski P and Thrower S J (1992) Modified atmosphere packaging of fruits and vegetables, sea-food and meat: State of the art. *ASEAN Food J* **7**: 127-136.

Ogawa Y, Morita K, Tanaka S, Setoguchi M and Thai C N (1998) Application of X-ray, CT for detection of physical foreign materials in foods. *Trans ASAE*, 41. 157-162.

Oh H S (2005) Fruit cleaning apparatus with dust collection system and fruit cleaning process using the same. Jeju Citrus Growers Agricultural Cooperative, 200504.

Park H J, Chinnan M S and Shewfelt R L (1994) Edible coating effects on storage life and quality of tomatoes. *J Food Sci* **59(3)**: 568-570.

Parson C S, Asen S and Stuart N W (1967) Contolled atmosphere storage of daffodil flowers. *Proceedings of Amer Soc Hort Sci* **90**: 506-514.

Peleg K and Ramraz Y (1975) Optimal sizing of citrus fruit. *Trans ASAE* **18**: 1035-1039.

Petersen A, Schlegel V, Hummel B, Cuendet L S, Geddes W F and Christensen C M (1956) Grain storage studies XXII. Influenec of oxygen and carbon dioxide concentration on mold growth and deterioration. *Cereal Chem* **33**: 53-66.

Peyton A J, Beck M S, Borges A R, Oliveria J E, de Lyon G M, Yu Z Z, Brown M W and Ferrerra J (1999) Development of electromagnetic tomography (EMT) for industrial applications. Part 1: sensor design and instrumentation. In: First World Congress on Industrial Process Tomography, Buxton, Greater Manchester, UK (pp. 306-312).

Primrose K M and Bolton G (2001) Applications of electrical process tomography in food processing in American Association of Chemical Engg. – Conference of Food Engg. – on the use of electrical resistance tomography in the food sector.

Prussia S C, Shewfelt R L, Bow man D W, Jordan J L, Craig W L, Fischer D F and Ashby B H (1989) Vibrational effects on produce deterioration rates. *Paper ASAE NO.* 89-6023.

Pruthi J S, Manan J K, Teotia M S, Setty S R, Eipeson W E, Saroja S and Chikkappaji K C (1984) Studies on the utilization of Kinnow and Malta oranges. *J Food Sci Tech* **21(3)**: 123-127.

Purvice A C (1983) Effects of seal packaging and ethylene removal in the sealed bags on the shelf life of mature green Japanese apricot. *J Japanese Soc for Hort Sci* **52(1)**: 85-92.

Randhawa J S, Dhatt A S, Rattanpal H S and Kaur H (1999) Storage behaviour of wax coated Kinnow mandarin at ambient conditions. *Proc Int Symposium on Citriculture.* NRC for Citrus, Nagpur, Maharashtra, India.

Rane A A (1983) *Economics of Agriculture.* pp 139-159. Atlantic Publishers and distributors, New Delhi, India.

Ranote P S and Bains G S (1982) Juice of Kinnow fruit. *Indian Food Packer* **36(5)**: 23-33.

Ranote P S, Saini S P S and Bawa A S (1993) Characteristics of Kinnow mandarin juice and beverages. *Indian Food Packer* **47(1)**: 11-13.

Rattanpal H S, Randhawa J S, Dhatt A S and Kaur H (1999) Studies on cold storage of wax coated Kinnow. *Proc Int Symposium on Citriculture.* NRC for Citrus, Nagpur, Maharashtra, India.

Raymond G, Mc Guire R G and Hagenmaier R D (2002) Shellac coatings for grapefruits that favour biological control of *Pecnillium digitatum* by candida oleophila. (Abstr available on sciencedirect.com).

Reddy R K, Shankaraiah V, Reddy Y N, Babu J D and Sharma P S (1999) Post-harvest treatments for more storage life of sweet oranges cv. Sathgudi at different storage conditions. *Proc Int Symposium on Citriculture.* NRC for Citrus, Nagpur, Maharashtra, India.

Rejimbal T R Jr and Bigler R E (1973) Cleaning efficiency of brush washers in citrus concentrate plants. *Proc Florida State Hort Soc* **85**: 254-257.

Saini S P S, Ranote P S, Mudahar G S, Bhatia B S and Singh K K (1985) Studies on processing of citrus fruits of Punjab. *5th Convention Food scientists and Technologists,* New Delhi.

Sandhu K S and Bhatia B S (1985) Physico- chemical changes during preparation of fruit juice concentrate. *J Food Sci Tech* **22(3)**: 202-206.

Sandhu K S, Bhatia B S and Shukla F C (1985) Physico-chemical changes during storage of Kinnow mandarin orange and pineapple juice concentrate. *J Food Sci Tech* **22(5)**: 342-346.

Sandhu S S, Dhillon B S and Singh S (1983) Post harvest application of G A and wrappers on the storage behaviour of Kinnow. *Indian Food Packer* **37(3)**: 65-71.

Santi R B and Carl M S (1988) Quality and shelf life of individually shrink wrapped peaches. *J Food Sci* **53(2)**: 519-522.

Sarmah U, Bains G S and Singh K K (1981) Studies on processing of Kinnow mandarin juice. *Pb Hort J* **21(1-2)**: 32-45.

Sethi V, Anand J C and Sexsena S K (1980) Kinnow orange in juice beverages making. *Indian Hort* **25(3)**: 13-15.

Sharma Amandeep and B S Ghuman (2008) Economic evaluation of on-farm waxing and grading unit for Kinnow fruit. *International J of Tropical Agriculture.* **26(1-2):** 221-223.

Sharma Amandeep and B S Ghuman (2007) Development and testing of prototype fpr on-farm waxing and grading of fresh Kinnow Mandarin. *International J of Tropical Agriculture.* **25(1-2):** 57-63.

Sharma Amandeep and B S Ghuman (2006) Status of Postharvest handling of Kinnow fruit in Punjab. *J Res Punjab Agric Univ.* **43(4):** 31-34.

Sharma G D and Kapur N S (1968) Role of skin coating on the transportation of perishable fruits. *Indian Food Packer* **22(3)**: 30-31.

Sharma R (1999) *Effect of Edible Coatings on Shelf-Life and Quality of Okra.* M.Tech. thesis, Punjab Agricultural University, Ludhiana, India.

Shyam M, Singh V and Singh R (1990) Design and development of potato grader. *AMA* **21(1)**: 40-42.

Singh A (1990) Experimental and analytical studies on Kinnows under controlled atmosphere, cold storage and ambient storage. Ph. D. dissertation, Punjab Agricultural University, Ludhiana, India.

Singh A, Singh Y and Batheja S (1992) Suitability of packaging for tomatoes. *J Food Sci* **29(6)**: 381-383.

Singh B J and Singh I S (1967) Potentiality of some citrus fruits for fresh processing industry. *Indian J Hort* **24**: 133-141.

Singh B P, Gupta A K and Chundawat B S (1973) Effect of various treatments on storage of Kinnow fruits. *Pb Hort J* **13(3-4)**: 161-164.

Singh D and Thakur A K (1999) Effect of harvesting and grading methods on storage loss of Kinnow. *Proc Int Symposium on Citriculture*. NRC for Citrus, Nagpur, Maharashtra, India .

Singh K, Mann S S and Chander M (1986) Proper storage of Kinnow fruits. *Indian J Hort* **31(3)**: 21-23.

Singh S and Naqvi S A M H (ed) (2001) *Citrus*. pp 27-31. International Book Distributing Co., Lucknow, India.

Singh S N (1981) Investigation on storage of Kinnow. *Pb Hort J* **21(1-2)**: 23-36.

Singh S, Shivankar V J, Ladaniya M S, Huchche A D, Shirgure P S, Marathe R A, Das A K and Wanjari V (2002) *Citrus 1001 Question – Answers*. Kalyani Publishers, Ludhiana, India.

Spivakovsky A and Dyachkov (1985) *Conveying Machines*. Afanasyev V (tr). pp 34-67. Mir Publishers, Moscow.

Srevens G N and Gale G E (1970) Investigations into the feasibility of semi-automatic quality inspection of fruit and vegetables. *J Agric Eng Res* **15(1)**: 52-64.

Srivastava K C and Bopaiah (1978) Prospects of growing Kinnow mandarin in the southern states. *Pb Hort J* **18(3-4)**: 139-141.

Sudheer K P, Chnadra P and Pal R K (1998) Modified atmosphere storage of fruits and vegetables using silicon membrane system: A feasible technology for India. *Agril Engg Today* **22**: 35-43.

Talasila P C, Chau K V and Brecht J K (1995) Design of modified atmosphere packages for fruits and vegetables. *ASAE Paper No 92-6020*. St. Joseph, Michigan, USA.

Thomas P, Dalal V B and Srivatava H C (1964) A comparison between the effects of wax and polyethylene coating on the storage behaviour of yellow limes. *Indian Food Packer* **1**:7-10

Thompson A K (1996) *Postharvest Technology of Fruits and Vegetables*. Blackwell Science Ltd, London, UK.

Tillett R D (1990) Image analysis for agriculture processes. Division Note DN 1585 Silsoe Research Institute.

Timmermans A J M (1998) Computer vision system for on-line sorting of pot plants based of learning techniques. *Acta Horticulture*. 421, 91-98.

Ting S V and Rouseff R L (1986) *Citrus Fruits and Their Products – Chemistry and Technology*. Marcel Dekker Inc, New York.

Toledo R, Steinberg M P and Nelson A I (1969) Heat of respiration of fresh produce as affected by controlled atmosphere. *J Food Sci* **34**: 261-264.

Torres J A (1994) *Edible films and coatings from proteins*. In: N S Hettiarachchy and G R Zeigler (eds) *Protein Functionality in Food Systems*, New York: Marcel Dekker, pp. 467-507.

Vedini K H (2001) *Agriculture Marketing Interventions and Innovations*. pp 330-337. National Institute of Agric Ext Mgt, Hyderabad, India.

Veraverbeke E A, Bruaene N, Oostveldt P and Nicolai B M (2001) Non destructive analysis of the wax layer of apple (Malus domestica Borkh.) by means of confocal laser scanning microscopy. *Planta* **213(4)**: 525-533 (Original not seen. Abstr in FSTA, Entry No 200112, 2001).

Wadowaski W F, Nagy S and Grierson W (ed) (1986) *Fresh Citrus Fruits*. AVI Publishing Co., Westport, CT.

Wagner A B and Sauls J (2004) Packingline Operations. (www. Extension-horticulture.tamu.edu/citrus.12292.htm)

Woodroof J G (ed) (1986) Fruit Harvesting Handling and storing in Commercial Fruit Processing. AVI Publishing Co., Westport, CT.

Yanai S T, Ishitani and Kojo T (1980) The effects of low O_2 atmosphere on growth of fungi. *Nippon Kogyo Gakkai-shi* **27**: 20-24 (Original not seen, Cited by Singh A (1990)

Experimental and analytical studies on Kinnows under controlled atmosphere, cold storage and ambient storage. Ph. D. dissertation, Punjab Agricultural University, Ludhiana, India).

Zagory D and Kader A A (1988) Modified atmosphere packaging of fresh produce. *Food Tech* **42**: 70-77.

INDEX